Parson RAMBINIZANDRY

MAHAFALE GRASSROOTS COMMUNITIES AND THE CONSERVATION OF FOREST RESOURCES

Parson RAMBINIZANDRY

MAHAFALE GRASSROOTS COMMUNITIES AND THE CONSERVATION OF FOREST RESOURCES

ScienciaScripts

Imprint

Cover image: www.ingimage.com

This book is a translation from the original published under ISBN 978-620-6-69878-4.

Publisher:
Sciencia Scripts
is a trademark of
Dodo Books Indian Ocean Ltd. and OmniScriptum S.R.L publishing group

120 High Road, East Finchley, London, N2 9ED, United Kingdom
Str. Armeneasca 28/1, office 1, Chisinau MD-2012, Republic of Moldova, Europe
Printed at: see last page
ISBN: 978-620-8-14513-2

MAHAFALE GRASSROOTS COMMUNITIES AND CONSERVATION OF FOREST RESOURCES

Parson RAMBINIZANDRY

ACKNOWLEDGEMENTS

This work could not have been completed without the invaluable help and collaboration of various people and institutions. I would therefore like to express my thanks and deepest gratitude to all those who helped me to complete this doctoral thesis.

We owe a very deep personal debt of gratitude to :

All the members of the Jury who agreed to sit and judge this research work;

Mrs Noeline RAMANDIMBIARISON, Thesis Director, Professor Emeritus. You have welcomed us with kindness despite your heavy obligations in professional, academic and family life. During the preparation of this book, you guided our research work in the right direction, particularly at each stage of the progress of our investigations;

Mr Jean Claude RAMANDIMBIARISON, Professor Emeritus, who assisted me and gave me advice in the elaboration of this work through his experiences. His comments were of great importance for the orientation of my studies;

Mr Rolland RAZAFINDRAIBE, Professor, who gave us invaluable advice throughout the preparation of this book, not forgetting his moral support;

Secondly, we must not forget all the teachers in the Sociology Department.

We would like to express our sincere gratitude to all the communities on the Mahafale plateau,

We would also like to thank the WWF team and the MNP for introducing us to the communities of the Mahafale plateau;

Finally, we would like to thank all the members of our family: our parents, our godfather, our brothers and our sisters. They supported us financially and morally to complete this thesis. To all of you, a thousand thanks.

CONTENTS

GENERAL INTRODUCTION

Since ratifying the CBD (Convention on Biological Diversity) in the late 1980s, Madagascar has been recognised as a country of great diversity; in terms of surface area, it is the only country with high concentrations of biological endemicity across different taxonomic levels. The extraordinary number of species (more than 12,000 plant species and more than 1,000 vertebrate species), and high levels of endemicity (more than 90 per cent of species are found nowhere else on earth) mean that the country's forests and biodiversity represent a unique and irreplaceable global public good, accounting for five per cent of the world's biodiversity in just 0.4 per cent of its land area[1] .

As a result, for more than a decade, Madagascar has had a contrasting public image worldwide, characterised by an exotic nature and exceptional biodiversity, with one of the highest rates of endemism in the world, and very heavy deforestation, a very rapid and catastrophic degradation of this rich and varied environment with, in addition, an annual blaze in most of the natural areas causing directly or indirectly the loss of considerable areas of forest in a desolate landscape of bare moors. This loss is spread over a large proportion of the area and is exacerbated by climate change, which is causing worrying erosion where inappropriate and degrading farming and grazing practices are rife. Added to this is a "galloping" demographic trend leading to land clearance, which in turn fuels deforestation and reinforces environmental degradation and soil erosion, leading to worrying rural poverty. Yet these farmers are part of our nation, as the Malagasy proverb says: *"Ny tanimbary fotaka, ny vola taratasy, ary ny omby fiompy fa ny Olona no Harena"*, literally the rice field is just mud, money is just paper, the zebu is just cattle, but it's the people who make up a country's resources.

Madagascar's forests have been seriously destroyed. Barely 10% of the natural forest has been preserved. Slash-and-burn agriculture and the massive use of charcoal and wood fuels have been particularly damaging to our natural resources. Yet these forests are important in more ways than one. They are the habitat of a large number of endemic plants and animals, and for the Malagasy people they are an essential resource with great potential, which must be properly managed and protected.

Thus, in 2003, at the Vth World Congress on National Parks in Durban, South Africa, known as the "Durban Vision", the Malagasy government, led by President *RAVALOMANANA*, undertook to triple Madagascar's coverage of protected areas to 6 million hectares, or to cover ten per cent of the country's surface area. This commitment on the part of the Malagasy State, caused by the triumph of liberal ideas, is combined with globalisation, which defines a new role for the State. The notion of the welfare state was no longer appropriate. The State, disengaged from all sectors, assumed its new role as coordinator and facilitator of economic and social development.

[1]World Bank: Project Document relating to a proposal for an additional IDA credit and a proposal for an additional grant from the Global Environment Facility Trust Fund to the Republic of Madagascar for the project to support the third phase of Madagascar's environmental programme, 2011.

Presentation of the study area

The South-West Region is home to a number of national parks with unique biodiversity (dry and thorny forests). *Tsimanampesotse* in the *Mahafale* plateau is one of these parks.

Located in the south-west region of Madagascar, the *Mahafale* region is home to the *Mahafale* camp. Most of its members are agro-pastoralists for whom the *Mahafale* plateau is valued as a resource for fodder, wood and food (food crops, gatherings, tubers, honey, etc.). The way of life of the local populations is based on a spatial and temporal balance of livestock farming and agricultural activities. More than 96% of the inhabitants make their living from farming and livestock rearing, although cultivated land accounts for no more than 20% of the total area of this landscape[2] . In this configuration, cattle rearing based on transhumance plays a key role among the *Mahafale*. However, farming is still at a very outdated stage for most farmers, and 95% of plantations do not use any fertiliser, whether mineral or organic[3] . At the same time, the livestock sector, particularly the zebu sector, has evolved very little and has remained at the contemplative stage, more cultural than economic.

The *Mahafale* region is increasingly affected by the phenomenon of desertification linked to climate change due to the irregular distribution of rainfall. For more than a decade, this has resulted in a decline in vegetation and soil degradation in cultivated areas, the worst of which are found along the coast (in the communes of *Beheloke*, *Itampolo* and *Salary* Sud)[4] . At 55%, the population is no longer able to meet its minimum energy requirements for upkeep[5] , and most of the population has become impoverished in recent years [Barraud, 2006].

Located in the South-West Region of Madagascar, the *Mahafale* zone alone covers a quarter of the region's total surface area. It is made up of the limestone plateau and is divided into three distinct geomorphological units that follow one another from west to east: the coastal plain, the limestone plateau and the basement or peneplain. The *Mahafale* limestone plateau covers an area of 12,500 km^2 .

[2] SIRSA 2004 and SAP 2010 data
[3] Agricultural Statistics Service, MAEP, 2006
[4] FOFIFA / INSTAT / Cornell - 2001
[5] In PROJET COGESFOR du PLATEAU MAHAFALE : RAPPORT D'ACTIVITES ANNUEL (MARS-DECEMBRE 2010)

Map 1: Location of our study area

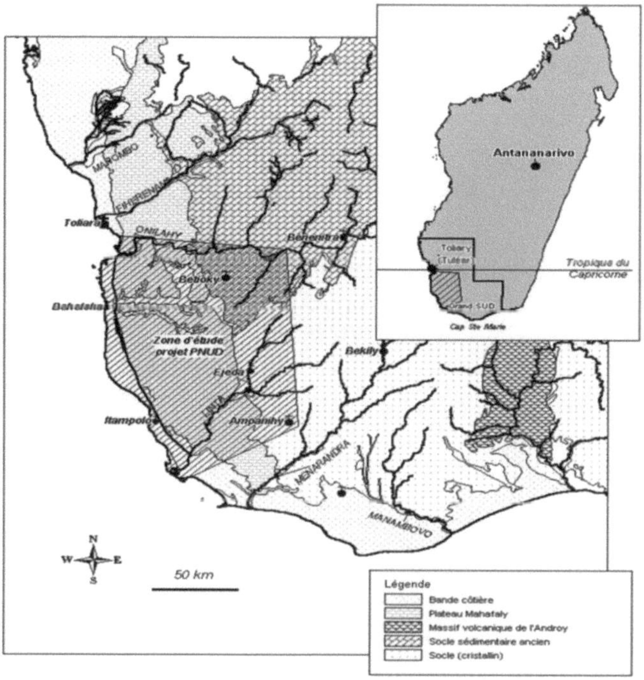

Source: WWF (COGESFOR, *TOLIARA*), 2013

The *Mahafale* landscape is bordered to the north by the *Tsimanampesotse* National Park. It encompasses 13 Communes who are aware of the degradation of the ecosystem of the *Mahafale* limestone plateau. What's more, WWF, through the COGESFOR project[6] , is beginning to take an interest in the richness of this *Mahafale* limestone plateau and, as a result, is offering support to grassroots communities to rationally manage its renewable natural resources through a transfer of management of these resources from the forestry administration, which is why the *Mahafale* plateau was chosen.

[6] A conservation and management project for Madagascan forest ecosystems initiated by WWF.

Map 2: Administrative map of the Communes making up the *Mahafale* landscape

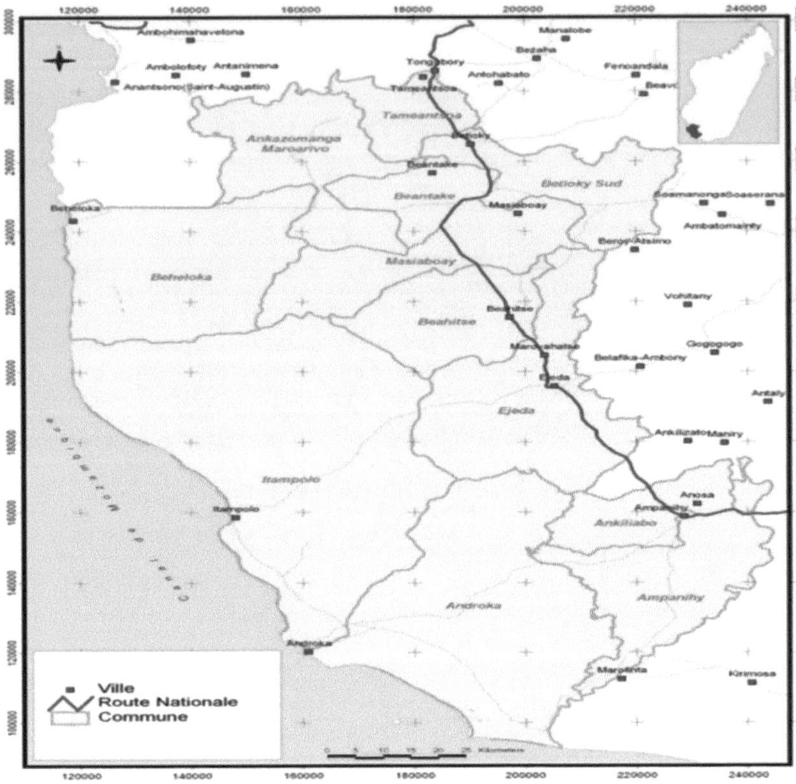

Source: WWF Toliara, 2013

Under these conditions, the landscape is home to a series of large forest blocks with a total area of up to 750,000 ha, equivalent to more than 43% of its total surface area. *Tsimanampesotse* National Park forms the core of the landscape. With a surface area of 207,000 ha, it accounts for 12% of the total surface area of this limestone plateau and is exposed to a variety of dangers, particularly those of man-made origin. The most characteristic of these are the slash-and-burn or *"hatsake"* farming practices and the transhumance practices, which tend to be long both in duration and distance. These practices lead to overgrazing, degradation of the plant cover and soil, and social conflicts between farmers and herders. This situation is exacerbated by the continuing increase in population numbers.

Faced with the ineffectiveness of the regulatory and legislative measures prohibiting land clearance and deforestation enacted by the forestry administration, or in the absence of adequate rules for transhumance, combined with the precarious living conditions in this region, these populations cannot help but resort to practices that do not respect the environment in order to survive. As a result, the entire *Tsimanampesotse* National Park and its outlying areas have been affected, and the inhabitants continue to live in persistent poverty.

In addition, the *Mahafale* sub-region is isolated, with a lack of suitable infrastructure for both inter-regional and inter-communal exchanges, reducing development opportunities with the outside world. This is compounded by food insecurity caused by chronic droughts that continue to intensify year after year. As a result, *Mahafale* communities are turning to coal production. This charcoal activity represents an alternative source of subsistence for farmers during lean periods.

Reason for choosing the subject

Our investigation focuses particularly on the *Mahafale* landscape around *Tsimanampesotse* Park. This delimitation is a scientific challenge, given the various geographical, sociological, psychological and political parameters involved.

Following the Malagasy government's commitment to the "Durban Vision", the size of the *Tsimanampesotse* National Park has increased to 207,000 ha from its original size of

17,520 ha. When it was managed by ANGAP from 3 October 2000, its surface area covered 43,200 ha. This means that 163,800 ha of community use rights are appropriated for the park. In addition, 328,410 ha were transferred to the communities to enclose the park so that it would not be subject to clearing by the local population. This means that there are insufficient natural resources for the use of the communities on the periphery of the park, and yet these communities are already weakened by the yoke of scarcity. They have been expropriated from their ancestral territories and find themselves at an impasse, using whatever they can find easily and quickly to support themselves. They are sinking into great destitution, the causes of which are as much social as they are geopolitical, economic or institutional.

Furthermore, the donors and non-governmental organisations (NGOs) working in the environmental field with the forestry administration adopt the conservationist trend, arguing that nature must be conserved. To achieve this, man must be ignored, as he is considered to be the enemy of nature. These forests have to be "sealed off" or protected, and governed by regulations, while the communities are driven out of their land altogether. They are marginalised. They can no longer freely use these resources without permission.

This raises the question of whether the management transfer initiated in 1996 has helped to conserve biodiversity and improve living conditions for local people.

This is why we have limited this research to the implementation of forest management transfers around Tsimanampesotse Park, but we will briefly discuss the implementation of a *Nosy Ve-Androka* marine protected area.

Aims of the study

Our theme is entitled "*Mahafale* grassroots communities and the conservation of forest resources". Community participation in conservation is analysed on the basis of the implementation of natural resource management transfers surrounding the *Tsimanampesotse* national park in the *Mahafale* plateau.

This study therefore aims to assess the scope of the participatory approach used by several NGOs, in particular the Deutsche Gesellschaft für Internationale Zusammenarbeit (GIZ) or

German Agency for International Development, SAGE, the World Wide Fund for Nature (WWF) or Global Environment Facility and the Madagascar National Park (MNP) working in the countryside, in the evaluation of the results of these non-governmental organisations (NGOs) working in the field of biodiversity conservation, but also in that of the grassroots communities (CoBa) managing the transferred resources, on the one hand, and their impact on the target beneficiaries, on the other. The NGOs have initiated a number of activities that are helping us to analyse the motivation of grassroots communities to take their future into their own hands.

Specifically, our study involves :

- to obtain information on the existing social organisation and its role in the production process in order to understand the prospects for the rural world;
- to understand community organisation and the role of gender in the protection of renewable natural resources, in order to assess village perceptions of what is meant by the environment and the purpose of the environment itself;
- to determine village perceptions of the opportunities/threats to the conservation of renewable natural resources in order to understand why this type of participation has emerged in the *Mahafale* landscape.

Expected results

The study will highlight the following points through conclusions/recommendations:

- perceptions of biodiversity conservation by grassroots communities or citizens ;
- the positive factors determining people's ability to adopt or change their behaviour with regard to the conservation of renewable resources;
- the aspirations of the Basa communities (CoBa) on the type of organisation and protection/conservation applied to curb the spread of land clearing.

Since the Rio Summit in 1992, the issue of economic and social development has been combined with that of the natural environment, i.e. the depletion of the ozone layer, global warming and the loss of biological diversity. This inclusion of the environmental dimension in development issues has enabled the human sciences to renew their original links with the natural sciences in this specific field, but although it is welcome, it is not without its questions. Indeed, one has to ask why it has taken so long for those involved in development to understand the obvious link between the economy, the social, in the cultural sense of the term, and the environment. In other words, why have we thought about the world without distinguishing between nature and culture?

Madagascar, and more specifically the *Mahafale* plateau, is an ideal place to reflect on all the issues that concern a large number of countries and that raise questions about the many disciplines and the relationships that man has with his environment and natural resources today.

Madagascar is stigmatised for its contrasts: the poverty of its people and the richness of its specific biodiversity. This richness is extremely important in terms of the endemicity of species, but it is also extremely important in terms of poverty, given that Madagascar is one of the poorest countries on the planet. To illustrate this contrast, poverty persists: according to a 2012 World Bank report, 92% of Madagascar's population was considered to be living in extreme poverty, while the natural endemism of species was estimated at 80%. This is why Madagascar is one of the world's biodiversity *"hot spots"*[7] . In Madagascar, talking about the environment means talking about biodiversity.

75% of the Malagasy population work in the rural sector, where 86% of all jobs are held, 60% of them by young people. The sector is characterised by its poor performance, reflected in its stagnant contribution of around 26% to GDP and its growth rate of 1.5%, well below the demographic rate [National Development Plan, 2015]. Various factors explain this situation: inadequate rural financing, low productivity, fragmentation of production, isolation of production areas, overexploitation of resources and insecurity.

This means that the welfare state can no longer satisfy all the needs or solve all the problems of its citizens. The centralised government is far removed from the people, and the hoped-for development is not taking place as expected.

Since the structural adjustment of 1980 and, above all, since the birth of the 3rd Republic, the State has gradually disengaged from both businesses and public institutions. Development is now the responsibility of economic operators, who are dispersed throughout the decentralised communities, which are becoming autonomous entities that form the basis of national development.

And yet, as the community suffers from poverty, a feeling of powerlessness sets in, marked by a half-hearted resignation. But a gradual disinterest in politics is also gaining ground, giving way to daily resourcefulness. The solution for the community is to plunder natural resources, whereas conservationists believe that nature must be conserved and, to do so, it must be done without man, man being considered the enemy of nature. By lobbying donors, the

[7] Very sensitive area in French.

conservationists thought they were putting Madagascar's natural resources under wraps as part of its ecological heritage, and the communities living in the forests would be driven out. Malagasy society is thus at an impasse. How can we mobilise people once again, give them a shared value and win the fight for development together?

Some say that development has to start from the bottom up or it won't happen. It therefore requires the active participation of everyone. But in the review of rural studies, No. 178, entitled "Development in Madagascar", Sophie Moreau talks about the dialogue of the deaf between NGOs and farmers:

> *"Despite the participatory approach advocated, the activities undertaken in many villages are not designed by the local people, but insofar as they fit in with the principles of environmental policy".*

This is why we want to study the participation of the grassroots community of the *Mahafale* landscape in the conservation/management of natural resources.

Approach method

To answer the questions set out below, we used a research method that enabled us to gather good quality information from the population concerned, and which helped us to gain a better understanding of the realities in this area. The methods used are set out below:

- **Sociological.** We need to adopt a dynamic categorical structuralism, postulating that the cultural elements transformed by the legal process are 'pre-constrained' in such a way that their reinterpretation is possible. process are 'pre-stressed' in such a way that their reinterpretation makes them more or less resistant to the change sought by public action. By observing phenomena of resistance, we can infer laws of transformation, a 'structure' that would explain why certain reinterpretations of endogenous law allow foreign models to be reappropriated, while others oppose them.
- **Psychosociology**: a special science that is dependent on and complementary to sociology. It is a method that studies social interaction and is therefore interested in the psychological conditions for the development of social groups.
- **Statistics**: set of mathematical methods, applied in sociology, which allow extrapolation and forecasting based on information obtained after processing data collected from a sample.

These methods are used to research and analyse the key concepts of this study, such as: the theoretical framework of participation in conservation; the conditions for grassroots community participation in the protection of natural resources; and the analysis and critique of grassroots community participation.

Techniques used :

We will therefore combine the three main search techniques, such as :

- Documentary analysis, in order to gain a better understanding of the data in our research field and of the reality to be studied, as well as to better define our research theme;
- information collected by means of an open-ended and semi-open-ended questionnaire;

- observation, to get a more objective view of the phenomenon under study and to verify what the respondents said, so that we are not content with mere abstractions or speculative studies.

1. Documentary analysis

It was carried out throughout the study, with a back-and-forth to gather additional information and bibliographical studies in the various documentation centres below: National Archives; Archive of the *Malagasy* Academy; Museum of Art and Archaeology of *Toliara*; Documentation Centre of the DEGS Faculty *of Antananarivo*, the Faculty of Letters and Human Sciences of *Antananarivo* and *Toliara*; University Library of *Antananarivo*; documentation in the library of the South-West Region; documentation in the library of the Ministry of the Environment and in the DREEF of *Toliara*; documentation from organisations working in the study area, such as WWF, MNP and GIZ; communal documentation, in particular communal development plans (PCD) integrating the village monograph of the study localities, the official gazette, magazines, national newspapers and dailies.

2. Gathering information in the field

Any hypothesis is only truc insofar as it can be verified by concrete quality information. Information has therefore been gathered in areas where there are grassroots communities managing forest and marine resources.

. **Sampling:** we conducted our study with a sample representative of the population as a whole. We opted for quota sampling in order to obtain the desired data from as many representative cases as possible. In addition, the pre-survey enabled us to select nine *Fokontany* according to whether they were economically poor, average or advanced in relation to the others. The people surveyed were taken at random from each of the categories defined.

The study site is the *Mahafale* plateau, home to three Districts such as *Ampanihy* Ouest, *Betioky* and *Toliara* II, in the South-West Region. The research was carried out in 20 of the 25 *Fokontany* that are home to the basic communities that manage forest resources. The total population of the *Mahafale* plateau is 219,042. If we took the extrapolated population numbers as a basis. We opted for a sample size of 4/1000, i.e. 800 individuals to be surveyed.

. **Living techniques:** we used a n interview questionnaire for the sample taken at population level. For members and non-members of CoBa, a focus group was organised to obtain more precise information on the conservation of natural resources in general and the nine research *Fokontany* in particular.

Our fieldwork lasted about eight (8) months. We divided our approach into two phases and two times: firstly, by directly approaching those in charge at commune level as well as those at the Direction Régionale de l'Environnement de l'Ecologie et de la Forêt (DREEF). After talking to the heads of the public authorities, we interviewed the heads of associations or NGOs working in the field of the environment (preservation of biodiversity). Finally, we conducted surveys in the *Fokontany* where the headquarters of each Communauté de Base (CoBa) are located. We used the quota method to select the population to be interviewed, by randomly choosing the households of members of the CoBa and the households of non-members (from among the 25 CoBa around *Tsimanampesotse* Park, which acts as the park's green belt). As a sample per

CoBa, we planned to survey 40 households, 20 of which were CoBa members and 20 non-members. They lived in the same *Fokontany*. We used the *Fokontany* register and randomly selected one household out of ten from those listed in the register, or, if necessary, directly on the spot.

The survey focused on habits and customs, production practices, household income, the sanitary infrastructure available to them (latrines, drinking water, etc.), their participation in biodiversity conservation activities, their behavioural changes, etc.

A questionnaire is drawn up for this purpose, and is analysed in quantitative form. If necessary, it will be adjusted to the field to measure the extent of certain elements (ideas, opinions or practices) that may have emerged from the discussion groups or interviews.

We therefore used subjective techniques such as in-depth interviews, life stories and focus groups.

Focus group or group discussion with targets

The subject of our study is a behavioural study. The focus group discussions (FGDs) enabled us to gain in-depth knowledge of the motivations behind the opinions expressed by the participants on the subject of the study. As a result, it was possible to identify the socio-cultural values underlying the responses expressed. In-depth interviews were also conducted with the public authorities (mayors, forest district chiefs, etc.), key people (members of the CoBa management committee, etc.) and project managers working in biodiversity conservation (WWF, ANGAP, *Tany Meva/SGP,* etc.).

These managers or key people also took part in the moderated discussion.

Being qualitative in nature, the results of the focus groups reflected the trends in the ideas and options of the groups of participants who reacted to them in relation to each other.

The target groups who took part in the focus groups were the same as in the quantitative study, i.e. men and women meeting the following criteria: living in the village, over 18 years of age, poor or non-poor, literate or non-literate.

Variables used to determine target groups

The study is aimed at homogeneous groups with the following basic variables:

- age: 18 to 45; 46 and over.

The 18 to 45 age group, or active adults, is made up of adults who are fully involved in various activities. This group appears to be the immediate successor to the older generation.

The 46+ age group, or seniors, encompasses the effective holders of economic and political power. As repositories of experience and wisdom, or opinion leaders, they have played an important role in determining a vision for the future and starting to implement it.

- socio-economic status (SES): poor/non-poor.

In this study, the poor were defined as those who are unable to achieve self-sufficiency: they go through a lean period; they are unable to send their children to school and are classed as destitute by the health centres.

On the other hand, the "non-poor" achieve self-sufficiency, are able to send children to school and make a financial contribution to health centres.

- level of education: literate/non-literate ;

The literate and the cultured take different intellectual approaches, which can only influence their vision of the future.

- sex.

Discussion groups of 6 to 10 participants were set up in each *Fokontany* where CoBa has its headquarters, selected at random from among CoBa members and non-members. In each *Fokontany*, we were able to organise two discussion groups. The groups were mixed, with as many men and women as possible. However, if the socio-cultural conditions prevailing in the locality were such as to restrict the free expression of one sex in the presence of the other, then for the two discussion groups, one would contain only women and the other only men.

The themes to be covered are as follows:

(i) knowledge, attitudes, practices and behaviour relating to the environment (sources of information and motivations for protection are among the subjects to be researched);

(ii) knowledge, attitudes, practices and behaviour relating to the conservation methods disseminated during awareness-raising (sources of information and motivations for adopting conservation or not, as well as knowledge of the link between the environment and the population are among the subjects to be researched);

(iii) CoBa knowledge, attitudes and practices with regard to renewable natural resource management techniques ;

(iv) attitudes towards the quality of services provided by CoBa management committees.

Discussions in the groups are facilitated by a moderator and recorded by a reporter. They are pre-analysed in the field during a debriefing between the moderator and the reporter. Their analysis is completed as and when data becomes available, whether in the field or in *Toliara*, and highlights the various trends relating to the themes and objectives under consideration. Discussions are based on a discussion guide drawn up for this purpose.

We therefore adopted a participatory method involving free and semi-structured interviews.

3. Observation in the field

In order to verify the statements made by the respondents, observations were made in the daily life of the communities. We forced ourselves to attend occasional events that may or may not be related to our research theme. We also accompanied some of the community patrol teams to observe what they were doing.

The concepts used

We will first try to clarify the concepts used in this research so that they can be read and understood in the same way throughout this book.

Briefly, a concept is an idea conceived by the mind. From a sociological point of view, a concept is a particular representation of a term to give it its own meaning and definition. Our research theme is made up of several concepts. This requires a definition to give us the same reading and mutual understanding.

Basic **Community Concept**

In its broadest sense, a community is a social group or collectivity with common property and interests (urban or rural commune; friendships; religious or professional association, etc.); it is an association based on the common ownership of the means of production and on complete or partial self-management.

The community (family, clan) can be found in all countries at the earliest stages of social development. It also existed in different forms in slave and feudal societies. The development of capitalism, by causing class differentiation within the community, led to its disintegration, and then its disappearance in industrialised countries.

Secondly, in its sociological sense, the community is a group whose members are bound together by participation in shared values. It has an internal solidarity that does not draw on explicit rules, but on deeper bonds than those of written law: the bonds of blood, those of tradition. This notion of community can be compared with Durkheim's opposition between organic solidarity and mechanical solidarity. But it was Ferdinand Tönnies (Community and Society) who constructed the ideal type of community which, after him, became frequently, if not always fruitfully, used in sociology.

The concept makes it possible to identify the difference between forms of sociability such as those of traditional society and industrial society. But it remains strongly marked by a critical evaluation of 'mechanical' societies and rationality, in favour of more 'natural' societies in which the social bond has no alienating exteriority.

In its strictest sense, this concept of a grassroots community is similar to what is meant by *"fokonolona"* in Malagasy (*Imerina*). The fokonolona is a traditional institution that played a major role in the old administrative organisation, at least in the *Merina* countries. It had an unwritten charter concerning its autonomy. Successive regimes in Madagascar have maintained it in different forms.

It should be pointed out that the *fokonolona* is purely "civil" in nature. It can neither set itself up as an operational structure, nor serve partisan interests linked to political challenges. The introduction of any form of subordination implicitly distorts the rules of *"fihavanana"*, which is its essence and its raison d'être. Its pluralism would be better used to inspire decisions than to impose them.

Under the transitional regime from 1972 to 1975, the Minister of the Interior set up the *fokonolona* for popular control of development. This *fokonolona* is identified by the implementation, through *dina* established in general assembly, of new attributions and responsibilities in terms of administration, management of its assets, mutual assistance, hygiene and public health, conciliation and arbitration in civil matters and development.

During the events of January to June 2002 in the country, the *fokonolona*, being a force of proposal, a force of interpellation and a counterweight to the authorities, demonstrated its lightning power, its determined presence and its mobilising capacity. Today, the *fokonolona* is called upon to play an important role in public safety, through the *dina be*[8] . This role has been strengthened by the creation of the "*andrimasom-pokonolona*" to combat the *malaso* phenomenon. Moreover, the *fokonolona* will be a local administrative unit for social, economic and educational activities. It will become a preferred area for rapid, harmonious and sustainable development.

In the context of the transfer of natural resource management, the grassroots community is a group of people responsible for ensuring the proper management of natural resources in a given area. In other words, the grassroots community is made up of any voluntary group of individuals united by the same interests and obeying the same rules of life. It includes the inhabitants of a hamlet, a village or a group of villages. The grassroots community has legal personality and operates like an NGO under the regulations in force. This organisation is governed by the GELOSE law or the GCF decree.

Conservation of natural resources

The Larousse dictionary defines conservation as the action of keeping something in good condition or preserving it from deterioration.

For the IUCN, conservation refers to the maintenance *in situ* of[9] ecosystems, natural and semi-natural habitats and viable populations of species in their natural environments and, in the case of domesticated or cultivated species (agro biodiversity[10]), in the environment where they have developed their distinctive properties.

Natural resources are defined as the set of units, whether renewable or not, that nature places at man's disposal, often as a heritage, and which he can use to enhance his well-being.

The conservation of natural resources must be implemented and accompanied by regulations. In Madagascar, when we talk about conservation, we are talking about a conservative conservation policy, in the strict sense of a specific model that separates the forest from society, ecology from everyday rural life, whereas the forest is a place inhabited by certain villagers. This would create a boundary between the villagers and the protected forests. There would no longer be any economic activities that people could undertake in the forests. Food insecurity would take its toll on farmers, while the human species must also be preserved in parallel with ecology.

Landscape concept

The Larousse dictionary defines a landscape as an area of land or countryside that offers an overall view.

---

[8] This is a collective agreement drawn up in the villages to control the invasion of stolen oxen in the rural communes of the Betioky and Ampanihy districts. This dina be is currently being adopted by the region's leaders to control insecurity in the entire region.

[9] In Lignes directrices pour l'application des catégories de gestion aux aires protégées, by Nigel Dubley in 2008: the Ecosystem is a dynamic complex of plant, animal and micro-organism communities and their non-living environment which, through their interaction, form a functional unit.

[10] Agrobiodiversity: Includes wild plants closely related to crops (wild relatives of cultivated varieties), cultivated plants (local breeds) and domesticated livestock varieties. Agrobiodiversity can be a protected area objective for crop relatives, traditional and endangered local breeds, particularly those dependent on traditional cropping practices; and/or traditional and endangered livestock breeds, especially if they are dependent on traditional cropping management systems that are compatible with "wild biodiversity".

According to the IUCN[11] (International Union for Conservation of Nature), the landscape is a contiguous area intermediate between "ecoregions" and "sites", comprising a mosaic of ecosystems (including those defined by human use) with a set of ecological, cultural and socio-economic characteristics distinct from adjacent areas.

An ecoregion is a relatively large terrestrial, freshwater or marine spatial unit containing geographically distinct assemblages of species, natural communities and environmental conditions. WWF's eco-regional approach and vision for biodiversity helps to define priority areas or priority landscapes for conservation.

The landscape approach takes into account the complexity of the relationships between ecological, social and economic functions, stakeholders, etc. The landscape integrates a wider range of conservation tools (protected areas, community management zones, sustainable production zones, agricultural zones, high conservation value zones, etc.).

Management transfer

It is a way of giving the community greater responsibility for managing natural resources. It involves a contract between the state, the municipality and the grassroots community to secure and manage renewable resources in the community's territory.

The State, represented by the forestry administration, may delegate the management of its forests to other public or private bodies (art 24 of law 97-017), in this case the said grassroots communities, by means of a management contract. This contract is governed by two laws, namely the GELOSE law (law 96 025 of 30/09/96) and the GCF decree (decree 2001-122 of 14/02/0).

Natural resources

The Robert dictionary defines resources as all the natural means available or available to a community. Also, RAMADE F. [1987] has specified that the term resource designates the entity in terms of energy and matter necessary for man to ensure his physiological functions or to fuel all his productive activities.

In a nutshell, natural resources are the set of units, renewable or not, that nature makes available to man, often by way of heritage, and which he can use to carry out his activities and increase his well-being.

Concept of participation

We will define participation as the act of joining or taking part in an activity. Alternatively, it can be the act of contributing to something in order to receive one's share. Participation also means taking part in decisions and/or controlling the application of those decisions.

---

[11] Founded in 1948, IUCN brings together States, government agencies and a wide range of non-governmental organizations in a unique global alliance: more than 1,000 members in some 160 countries. As a union, IUCN's mission is to influence, encourage and assist societies throughout the world to conserve the integrity and diversity of nature and to ensure that any use of natural resources is equitable and ecologically sustainable. To safeguard natural resources at local, regional and global levels, IUCN relies on its members, networks and partners, building their capacity and supporting global alliances.

The *Mahafale* landscape includes a protected area. This is the *Tsimanampesotse* National Park. The park was the subject of a declaration made by the then President of Madagascar in 2003 in Durban, South Africa, at the Vème World Congress on National Parks, who pledged to triple the surface area of protected areas in Madagascar. *Tsimanampesotse* National Park is one of the protected areas to be extended. Its new boundaries now extend to 240,000 ha, covering the Communes of *Beheloke*, *Beahitse* and *Itampolo*. Previously, the total surface area of the *Tsimanampesotse* park was 43,200 ha, entirely within the rural Commune of *Beheloke*. Since 2006, agents from the National Association for the Management of Protected Areas (ANGAP) have been working in the park, raising awareness of the need for conservation and sustainable management of natural resources, with a view to implementing this in the outlying areas of the park as a solution to the pressures. The aim of this extension is to enhance the potential attractiveness of the fauna and flora and to conserve the area's rich biodiversity.

This extension will have repercussions on the activities of the farmers around the park, as the land and production sites will also diminish.

However, the production method used is still generally very rustic and based on slash-and-burn. This is one of the main reasons for the loss of forest cover. This technique is used to conquer new land, but also to find more arable land suitable for cultivation. This situation is exacerbated by the continuing increase in population numbers.

The second threat to this ecosystem is the manufacture of charcoal. Species from the natural forests are used for this charcoal production, to provide a source of income for the local population. Finally, like all Malagasy societies, the *Mahafale* are major livestock breeders. Contemplative" cattle rearing is a symbol of wealth and has a very important socio-cultural value. The uncontrolled roaming of cattle, goats and sheep in the forests in search of fodder disrupts the structure of the forest and the development of regeneration, weakening the ecosystem and threatening the sustainability of the forests. It is also responsible for the spread of invasive species into the heart of the forest, and is another threat to the structure and regeneration of natural forests.

The way forest resources are managed today is different from what it used to be. Today's managers are under contract with the State, whereas in the past the *Mahafale* lineage society contracted with the ancestors or spirits through the *Ombiasy* and the *mpitan-kazomanga.* The latter manage forest resources using specific oral prescriptions known *as "lilindraza".* Consequently, this "traditional management" of the forests did not require a minimum of foreign management skills, which is nothing more than "management theorised" by Western experts. But are the communities of the *Mahafale* plateau, most of whom are illiterate, actively participating in this co-management of natural resources?

Tsimanampesotse National Park and its outlying areas are occupied by communities. These communities are obliged to protect renewable natural resources, even though the majority of the population continues to live in chronic poverty.

But what is motivating communities to preserve these renewable resources, when they are in the depths of poverty and have exploitable resources only in the *Mahafale* plateau?

So will communities be able to preserve these resources in the way they expect?

This study will be undertaken diachronically and synchronically with a dual objective: to evaluate the activities of grassroots communities and to understand the reality of the rural environment.

Research hypotheses

The implementation of forest resource management transfers around the *Tsimanampesotse* Park as a belt prompted us to contribute to this study to evaluate the participation of communities in preserving their resources. Our research theme is therefore entitled "*Mahafale* grassroots communities and the conservation of natural resources", but we have limited our field of study to the implementation of forest management transfers around *Tsimanampesotse* Park and to the establishment of marine protected areas from *Nosy ve to Androka.*

We opt for three hypotheses which we will try to verify in the last two parts of our book.

Assumption 1

Involving local communities in the management of renewable natural resources is very fashionable in Madagascar. At present, however, it seems to be the result of initiatives that are largely confined, if not discreet, and generally implemented under the impetus of donors.

Some believe that the involvement of these communities does not constitute a radical change in their practices and uses of forest resources. This does not mean that government bodies are not aware of forest/local community relations. Generally speaking, the practical introduction of forestry legislation relating to community involvement (secure local management [Gélose] and/or contract-based forest management [GCF]) still regards local communities as mere forest users. This probably explains the fundamental orientation of forest conservation projects towards limiting rather than strengthening acquired rights relating to forests.

Thus, the community-based approach planned to be applied in the context of natural resource conservation projects often seems to be aimed at replacing the local use of forest products and/or replacing economic dependence on forests by promoting improved agriculture or other economic alternatives such as ecotourism. However, the climatic conditions that prevail on the *Mahafale* plateau do not allow communities to adopt this improved agriculture. Involving local people in practical management tasks such as monitoring is not necessarily a meaningful form of participation.

Moreover, the form of participation in the conservation o f natural resources is different from that of community work. The conservation of natural resources requires the participation of all people, as well as a minimum knowledge of the forest ecosystem. So managing or protecting biodiversity is a job for people with solid skills and knowledge of the forest ecosystem. These people are called forestry technicians. However, the communities living near these forest resources do not have many instructions. In fact, according to the 1996 Human Development Report, Madagascar's capacity shortage index was 0.353 in 1993.

The educational level of the population of the *Mahafale* plateau is very low. For example, 85%[12] of the population of the capital of the two Districts (*Ampanihy* and *Betioky*) are illiterate. This

[12] PCD for the Rural Commune of Ampanihy, renewed in December 2008; For the Rural Commune of Betioky sud: the illiteracy rate in the town is around 75% and in the countryside reaches 90 to 95%.

situation has repercussions on the population's perception/representation of what is known as environmental conservation, even though the areas of forest in these two districts are currently very small. However, the donors are saying that the local communities need to be given a sense of responsibility by protecting the resources around them so that they no longer clear land or commit robberies, even though they are not technicians. But will they actually participate in the management of these natural resources?

The intellectual poverty of the Malagasy community prevents people from participating: they cannot take part in the political, social or economic life of their society because they have intellectual limitations, given their level o f education.

Thus, the first hypothesis is that **the level of participation of the grassroots community depends on its level of education, and non-participants at the lower end of the scale do not want to get involved in matters they consider obscure and dangerous.** We will therefore try to verify this hypothesis from the section dealing with the incapacity of the poor in part two.

Hypothesis 2

Each village has its own territory called a "*faritany*". Their boundaries are marked by trees, grasses, stones or other easily identifiable physical objects known as "*vorovoro*[13] ". The *faritany* is a mystical space where the sacred is strongly present, because it is in this sociological space that the dwellings of the deceased "*aritse*", the tombs and the sacrificial poles "*hazomanga*" are accumulated.

The *Mahafale* plateau is dotted with sacred forests that are respected by the communities. "*Alan-draza*" (ancestral forests) and "*ala faly*" (taboo or sacred forests) are the two names given to the forests on the *Mahafale* plateau. The ancestors decided that part of each forest was sacred or taboo. They are burial places that cannot be desecrated, but also places that must be preserved because they contain a wealth of fauna and flora. The wealth of sacred forests serves as a reserve in times of crisis: illness (medicinal plants), famine (wild yams, hedgehogs, etc.), disaster (timber for important community sites) and for building coffins for the dead. The forest inhabited by the spirit known as *Tambahoake* is also considered a taboo forest, as it is from here that the healers gather the various plants and roots to heal and protect their clients.

The forests have been handed down by ancestors since ancient times, which most communities are unable to define exactly, but which other communities date back to the 19th century, during the period of royalty, before colonisation.

In the concept of secure local management (Gélose), community conservation areas extend beyond resources (forests, lakes, etc.) and include part of the village land governed by the communities. It has to be said that community sites would only be eligible as community conservation areas if the value of their biodiversity, associated with their culture, is significant. Consequently, the social organisation of the *Mahafale* plateau is disrupted by external factors such as technicians or development projects.

However, if we look at it from the outside, it seems that the management of natural resources under the Agar or GCF contract is going in the expected direction. But is this new management

[13] These are clear markings that act as traditional boundaries, often temporary, to protect reserved areas.

strategy, standardised by experts from the international community, the right way to manage the natural resources of the *Mahafale* plateau?

Thus, the second hypothesis that we would like to verify is **that traditional social organisation and customs ensure the stability and success of management transfers in the *Mahafale* Plateau.** To this end, we will criticise the scope and limits of the natural resource management approach linked to the policy pursued by each stakeholder in the *Mahafale* Plateau in the final section.

Assumption 3

Local management of renewable natural resources by means of a management contract is a new approach that technicians and financial backers want to put in place throughout the island in order to achieve the Durban Vision: to increase the surface area of protected areas, and therefore of priority zones for protection. This means that the inhabitants will have to leave these 6 million hectares. But why will they have to leave the areas where they used to live?

At present, the aspiration of the "greens" to interfere in the management of Madagascar's renewable resources has been fulfilled. The transfer of management applied in Madagascar is modelled on the management of the colonial period, even though Madagascar is one of the countries in the world with the oldest references in terms of conservation policy. Colonial forestry policy was clearly based on the exclusion and repression of populations. Certain conservationist NGOs, in order to achieve their objectives, could take up the exclusion and repression of the forest dwellers who have lived there for a long time. Inhabitants, whether inside or on the periphery of these resources, will be excluded from their environments.

However, on the *Mahafale* plateau, the communities are suffering from a long drought and insufficient arable land, despite living very close to the park. They have adopted other solutions, which do not satisfy the donors, to survive: the plundering of turtles and the exploitation of forest resources (rosewood, etc.).

Thus, the creation of protected areas alone does not guarantee the sustainability of the biodiversity they contain. The conservation of biodiversity in protected areas depends on two factors: on the one hand, the reality of the State and its ability to enforce the rules it lays down [Bertrand, 2006]; on the other, the reaction of the resident populations who have long occupied these areas and lived off these resources and this biodiversity.

Therefore, the final hypothesis to be verified is that **the management transfers of the *Mahafale* Plateau do not achieve the objectives of sustainable management because of the reduction in the use rights of the communities, the geographical situation and climate change, which weigh heavily on their land**. In the sub-chapter dealing with the future of the forest ecosystem of the *Mahafale* Plateau, we will attempt to verify this last hypothesis.

Field approach

The visit to the site took place in 3 stages:

- February 2014 to May 2014: a survey was undertaken on the basis of questionnaires concerning communities' motivation to preserve and manage renewable resources but also concerning monographic data on the *Mahafale* landscape.

- June 2014 to July 2014: in-depth semi-structured interviews were conducted with *Fokontany* and Communes officials and forestry administration officials.

- From August 2014 to September 2014, focus groups were held with member communities of each village association of renewable resource managers, as well as with those who are not members of the resource managers, and finally with the management committees of each village where there is an association.

Table 1: Summary table of people surveyed

Features	Number	Type
Local managers	84	*Fokontany* and Communes officials, politicians, civil servants. senior and middle management, *V.O.I* presidents[14] by village
Affluent population	58	Notables, large farmers, transporters, collectors, elected representatives, clerics
Employed population	36	Employees of environmental NGOs, teachers, micro-finance employees
Working population	502	*FRAM*, population belonging to the *VOI*, population not belonging to the *VOI*
Population outside villages where management transfers exist	120	Visitors from each village inhabitant of the *VOI*, transhumants (and herdsmen), migrants from the village
TOTAL	**800**	

Source: Author, 2014

Finally, we conducted **6** focus groups, including 2 *VOI* members, 2 non-members and 2 notables.

Data processing method

The questionnaires completed by the interviewers are validated as soon as they are received. The questionnaires are then entered and processed in SPHINX 4.5 software. The data is summarised in the form of graphs for quantitative data and reported 'verbatim' for qualitative information, in order to respect the spirit of the author.

Limits

This research cannot claim to provide an absolute approach to biodiversity conservation. It is concerned only with the conservation and/or management of renewable natural resources around a national park, and specifically with the issue of the conservation of forest ecosystems around *Tsimanampesotse* National Park, to be promoted as part of the COGESFOR project under WWF supervision.

Nevertheless, the spirit of this research could inspire interventions in other, more diversified and broader areas. Despite this, our research was limited by the lack of statistical data available from the relevant administrative departments in the area.

What's more, we made a field trip during the hot season, when the *malaso* attack increases. They rob anyone they come across. Consequently, disturbed by this insecurity, as our fieldwork is none other than the home of these *malaso*, we were forced not to stay long in a few villages

[14] VOI: Vondron'Olona Ifotony

around the *Tsimanampesotse* park to observe the behaviour of the *Mahafale* villagers in relation to the protection and management of the transferred forests. What's more, the villagers are highly mobile, moving from one village to another according to rumours that the *Malaso* are going to attack their villages. The inhabitants flee and this will have an impact on the size of our survey sample.

Finally, for economic reasons, we have not been able to include transcripts of the interviews we conducted during our fieldwork.

Document outline

The thesis is presented in three parts, outlining the approach taken and summarising and discussing the main results.

The first part will therefore focus on the theoretical framework of community participation in conservation. In this part, we will discover the social structures of the *Mahafale* communities. Consequently, we will identify the physical and human environments, along with the techniques and livelihoods of the population, in the first chapter. Then, in the second chapter, we will discuss the conservation of these natural resources. Finally, we will analyse the process of village participation in the biodiversity conservation programme through the transfer of natural resource management: the actors and the approaches recommended in this process of implementation. This analysis of participation will be followed by a theoretical and descriptive approach to our theme and our fieldwork, in order to gain a better understanding of the *Mahafale* plateau.

The second part will be devoted to the conditions for citizen participation in the conservation of forest resources. In this part, and more specifically in the fourth chapter, we will attempt to explain the participation of stakeholders in the conservation of natural resources. The fifth chapter then sets out the framework for understanding the conditions for community participation in the conservation of forest and marine ecosystems. Finally, we will describe the accompanying measures that have been put in place as alternatives to conservation.

In the third part of our thesis, we will attempt to analyse and critique the scope and limitations of the participatory approach and of Agar and GCF in the seventh chapter. First, we will analyse the implications and limitations of the legal instruments. Secondly, we will criticise the scope and limitations of the policies pursued by the stakeholders.

In the final chapter, we shall conduct a general discussion of the various results by analysing the future of the transfer of natural resource management in the *Mahafale* Plateau and in Madagascar. Next, we will look at the trend in forest management, and whether the government will adopt a policy of sustainable management and development or a policy of preserving natural resources. Finally, recommendations will be presented, taking into account the various stakeholders.

Part One:

Theoretical framework for participation in conservation

We will attempt to take a theoretical and descriptive approach to our subject and our terrain, in order to gain a better understanding of the *Mahafale* plateau. In the first chapter, we will therefore describe the physical and human environments, together with the techniques and livelihoods of the local population.

And then we're going to take a theoretical look at the conservation of natural resources and participation.

Chapter 1: The Mahafale community and social structures

According to specialists, development is a set of social transformations that make autonomous, self-sustaining economic growth possible. At present, development follows a linear trajectory in which so-called "developing" societies are forced to "catch up" with the developed countries, abandoning their own identity in order to imitate them. This development was primarily conceived on the basis of technical innovations that had to be disseminated to the target societies. In a way, technical progress was granted to populations that, more often than not, had not asked for it. As a result, the vertical nature of the "support" was not thought to be a problem. It has only been found that this type of development destroys societies' capacity for innovation, their traditions and therefore their social capital.

Sustainable human development is another form of development that emphasises this community identity. It is the development of man by man and for man, in which man is the principal actor.

However, different parameters influence the participation of community members.

1.1 Profile of the *Mahafale* community

We are going to try and take a descriptive approach to our terrain, in order to gain a better understanding of the *Mahafale* plateau.

And then we'll look at the physical and human environments, with the techniques and livelihoods of the population.

1.1.1 Geographical location

1.1.1.1 Physical geography

The research was carried out in the heart of the *Mahafale* landscape[15] . This encompasses the *Mahafale* karst plateau, which is made up of three adjacent Districts, such as *Betioky* sud, *Ampanihy* and *Toliara* II, which are part of the *Atsimo Andrefana* Region, Province of *Toliara*. This part of the region is historically known as the place where the *Antanosy* and *Antandroy*

---

[15] The study area is made up of 13 communes (17,442 km²), home to more than 220,000 inhabitants (more than 70% of whom are agropastoralists) divided into 285 Fokontany.

populations migrated. Today, the Antanosy population is steadily expanding and settling in the various parts of the region.

Map 3: Map of the districts delimiting our study area

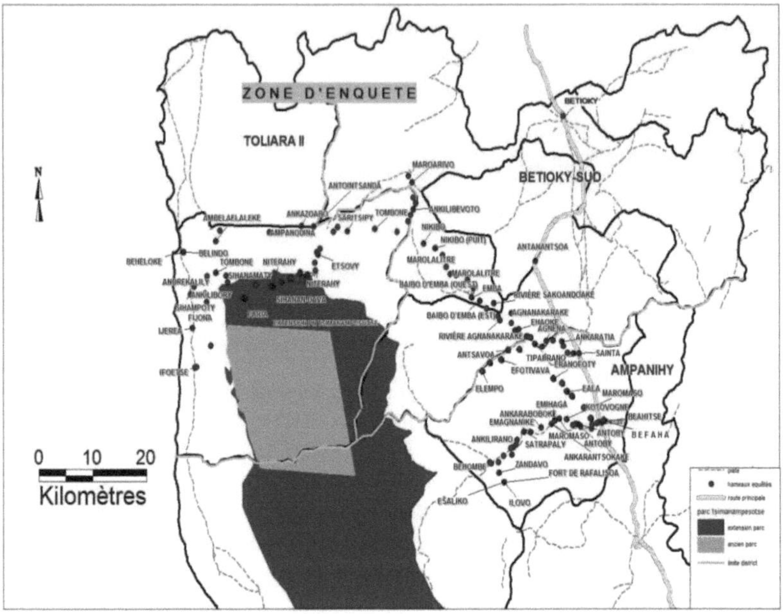

Source: Alexio Clovis, Toliara 2014

Located 160 km[16] from the town of *Toliara*, the *Mahafale* zone begins 95 km from the RN10: *Betioky* is the District capital. However, because of its access roads, it is landlocked and condemned to stagnation, even in the agricultural sector.

The *Mahafale* area alone covers a quarter of the Region's total surface area. It is made up of the limestone plateau and is divided into three distinct geomorphological units that follow one another from west to east: the coastal plain, the limestone plateau and the basement or peneplain.

The *Mahafale* limestone plateau covers an area of 12,500 km2. Its boundaries are as follows:

- to the north by the Rural Commune of *Tongobory* (*Onilahy* river),
- to the south by the Cap Sainte Marie Special Reserve,
- to the east by route nationale no. 10,
- to the West through the Mozambique Channel.

---

[16] 67 km along the RN7 to Andranovory: a branch from Toliara Antananarivo and Toliara Fort-Dauphin

1.1.1.2 Relief

The morphology of the region is characterised by three main zones distributed from east to west, stretching from north to south.

To the west, there is a continuous coastal plain varying in width from 1.5 to 15 km. This area is covered by quaternary sands interspersed with alluvial deposits and temporary pools. The *Tsimanampesotse* lake, 15 kilometres long, has been classified as a Ramsar site since 19 February 1998 and is part of the national park of the same name.

The lake is bordered to the east by the "*Mahafale* Plateau", a nummulitic limestone plateau that covers an area of 10,000 square kilometres and is generally between 100 and 200 metres above sea level. It also forms a large part of the *Tsimanampesotse* National Park. It is uninhabited and displays a variety of karstic features [Raunet 1996] :

- There are around a hundred deep sinkholes (40 to 100 metres deep) in the north-west section,
- An area of clay-bottomed sinkholes in the central and eastern parts,
- Depressions with decalcification clays, as well as dry corridors and valleys, evidence of a fossil hydrographic network, in the eastern part [Raunct 1996].

The third zone, to the east of the limestone plateau, is made up of a mosaic of limestone outcrops and depressions covered by thick Pliocene sandy deposits, often referred to as "red sands". The red sands give rise to tropical ferruginous soils with a rather acidic pH ("siliceous red earth"), sometimes mixed with ferralitic decalcification soils; or, less frequently, to more fertile hydromorphic soils in low-lying areas.

This plateau is generally covered by a thick limestone crust that only exceptionally reveals the rock in place. Slightly inclined towards the sea, it is cut by several valleys, both present and fossilised.

From north to south:

- the *Onilahy* valley, which is steep and particularly winding;
- the *Itombona* corridor, a rectilinear fossil valley running east-west and extending northwards into the *Ankazomanga* basin. The whole area is filled in by continental Neogene overlaid by outwash of continental red sands;
- the fossil valley of *Ilempo*, also filled in by red sands, forms a WNW-SEE corridor which, starting from *Behomby*, opens out to the south of the *Itomboina* corridor;
- The *Linta* valley cuts into the plateau in a SSW/NNE direction. As such, it is considered to be a river of the Deep South, unlike the River *Onilahy*. The *Linta* is completely lost underground as it enters the limestone terrain to the south of the town of *Ejeda*, so that it only flows for a few days a year in its lower reaches, which are completely filled in with sand and clay (wadi system),
- the *Sorombe* corridor, which joins the *Linta* in an east-west direction.

These valleys divide the *Mahafale* plateau into compartments from north to south:

✓ the northern plateau, between the *Onilahy* valley and the *Itombona* corridor;

- ✓ the large *Behomby* plateau, between the *Itombona* corridor and the *Linta* ;
- ✓ the *Ranomasy* plateau, between *Linta* and the *Sorombe* corridor;
- ✓ the *Ambovo* plateau, to the south, between the *Linta* and *Menarandra* rivers.

Map 4: Morphological subdivisions of the *Mahafale* landscape

Source: Aurouze, 1957

1.1.1.3 The floor

The soils in our study area are divided into three main varieties from west to east:

- To the west is a continuous coastal plain, covered by Quaternary sands interspersed with alluvial deposits and temporary pools;
- In the heart of the limestone plateau, decalcification clays give rise to ferralite-type soils ("red earth on limestone shell"), which are well-structured and have a neutral or basic pH;
- The eastern limestone plateau is made up of a mosaic of limestone outcrops and depressions covered by thick Pliocene sandy deposits, often referred to as "red sands". The red sands give rise to tropical ferruginous soils with a rather acidic pH ("siliceous red earth"), sometimes mixed with ferralitic decalcification soils; or, less frequently, to more fertile hydromorphic soils in low-lying areas.

1.1.1.4 The climate

The area's climate is marked by two seasons: a hot, rainy one from November to April, and a dry one from May to September. However, the local population differentiates between three main seasons:

- a hot, humid season (*asara*) from December to March,
- a cool dry season (*asotry*) from April to July,
- a hot dry season (*faosa*) from August to November.

Average annual rainfall varies from west to east across the plateau. It is around 300 to 350 mm in the coastal zone, 400 to 450 mm on the *Mahafale* plateau, and 500 to 600 mm in the savannah zone. The average annual temperature is 22.7°C. The hottest months are between December and March, with an average temperature of 24.7°C; the coolest months are between June and August, with an average temperature of 18.8°C.

Low rainfall, high evapotranspiration and the limestone nature of the bedrock contribute to the scarcity of water resources in the area. Water points are rare and most dry up during the dry season. The two maps below show variations in rainfall and temperature in the *Atsimo Andrefana* Region.

<u>Map 5</u>: Annual map of isohyets and isotherms in the South-West region

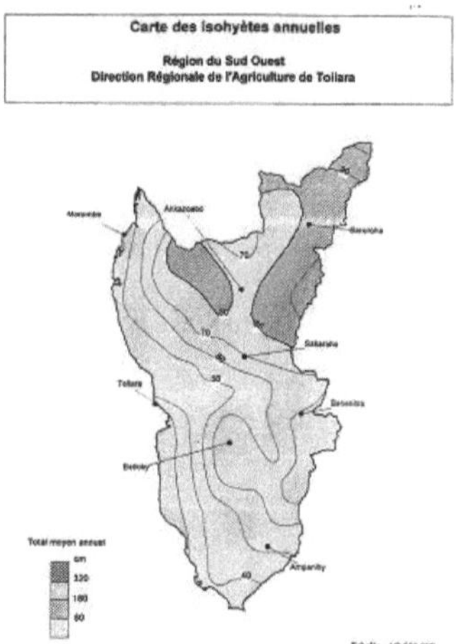

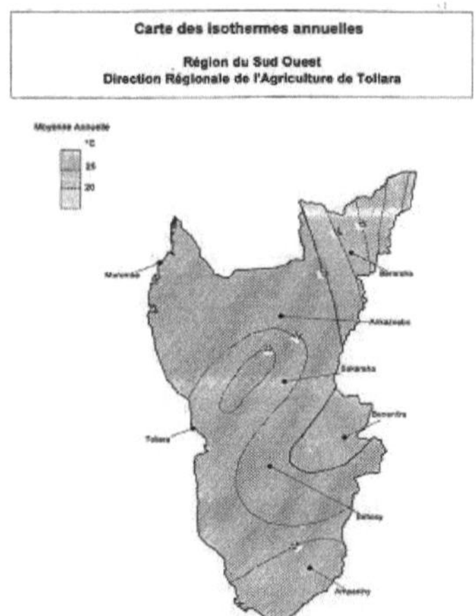

<u>Source</u>: UPDR, 2003

The *Mahafale* region is increasingly affected by the phenomenon of desertification linked to climate change due to the irregular distribution of rainfall.

1.1.1.5 Vegetation

The plateau is home to a series of large forest blocks with a total area of up to 750,000 ha, equivalent to more than 43% of its total surface area. The *Tsimanampesotse* National Park[17] forms the hard core of the landscape; its surface area of 207,000 ha corresponds to 12% of the total surface area of this limestone plateau.

The area belongs to the western phytogeographical domain, south-west sector [Humbert & Cours Darne 1965] and to the low altitude western ecofloristic zone [Rajeriarison & *Faramalala* 1999]. The vegetation is characterised by the dominance of xerophytic species [ANGAP, 2001], particularly on the limestone plateau.

Plant formations vary according to edaphic conditions (red sand and white sand).

In general, the following sequence is noted, from West to East:

- a coastal forest on the coastal plain, on sandy soils. Heavily degraded, it is known as xerophilous *samata* thicket or Euphorbia stenoclada (Euphorbiaceae),
- herbaceous plant groups on halomorphic marshy soils and woody plant groups on halomorphic soils, in lake or flood depressions (Lake *Tsimanampesotse* and salt ponds),
- a dense dry deciduous forest on the edges of the limestone plateau, in clay dolines and on red sand. It has an upper stratum reaching 10 to 12 m in height and is generally dominated by Didierea madagascariensis (Didieraceae),
- xerophytic thickets of varying density and height, also known as xerophytic bush, often dominated by Alluaudia comosa (Didieraceae) and coralliform euphorbias. These thickets are a form of adaptation of the dense dry deciduous forest to the difficult water conditions of the limestone plateau,
- a wooded savannah with Heteropogon contortus (Poaceae), dominated by Poupartia sp. (*sakoa*), and Tamarindus sp. (*kily*). It is probably anthropogenic and occurs on red sand to the east of the limestone plateau.

1.1.1.6 Hydrology [GUYOT L., 2002]

Battistini [1964] distinguishes between different basins depending on the organisation of the flow in the extreme south:

- exoreic basins, normally drained towards the sea,
- closed basins with endoreic drainage,

---

[17] History of the Park

Before 1960, Lake *Tsimanampesotse* and its surroundings were managed by the colonial administration, which created a 17,520 hectare "Integral Nature Reserve" (RNI) there in 1927. *Tsimanampesotse* was the tenth Integral Reserve created in Madagascar and the first protected area in the province of Toliara.

After the advent of the First Republic, the reserve was extended to 43,200 hectares and came under the authority of the Ministry of Water and Forests, becoming a category II protected area or Nature Park in June 1966 (decree 66-242), then a National Park in 2002 (decree no. 2002-797 creating the *Tsimanampesotse* National Park). Lake *Tsimanampesotse* was also classified as a RAMSAR site (wetland of international importance) by decree 98-003 of 19/02/98. A further extension in 2007 brought the lake to its current size of 203,400 hectares (see map 8 for the current boundaries).

- areal areas with no organised flow.
-

Map 6: Hydrography of the "Grand Sud" region

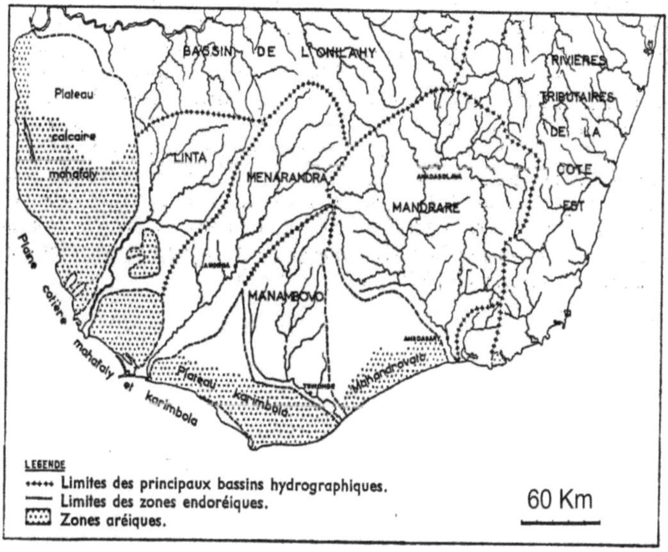

Source: Battistini 1964

The former correspond essentially to outcrops of the Precambrian basement and rare sections of the surrounding Neogene plateaux. In the Precambrian terrain, the river system is very dense and highly branched, underlining the structural layout of the Precambrian series. These exoreic expanses form the catchment areas of the major rivers, which receive most of their tributaries before entering the sedimentary zone, which they pass through "like strangers".

Bordering the study area to the north and south, the *Onilahy* and *Linta* rivers are of this type.

1.1.1.6.1 The *Onilahy* river

Onilahy is the geographical limit of the *Mahafale* landscape. It belongs to the south-western hydrological domain [Aldegheri, 1967 *in* Salomon 1986], i.e. it is subject to a tropical regime with two clearly defined seasons, with rainfall of between 500 and 800 mm/year. The river flows in a general east-west direction. It drains a basin of around 32,000km^2 [Chaperon et al. 1993], the upper part of which, where the hydrographic network is most developed, is located in the bedrock.

The *Linta* River

The *Linta*, which flows from north-north-east to south-south-west, is a river in the *Mahafale* landscape. Its basin is much more limited than that of the *Onilahy* (5800km^2) and extends over a less well-watered area. The *Linta* is therefore much less well supplied. Even at the *Ejeda* gauging station on the edge of the bedrock, surface run-off only occurs between November and

February. They are quite exceptional downstream, where the river is rapidly lost in the karstic network, so there can only be surface flow during very heavy rainfall upstream.

1.1.1.6.2 The areal zone of the plateau

The *Mahafale* limestone plateau currently forms a vast area with no organised permanent drainage, either because rainwater infiltrates it on site (lapiazé limestone in the pothole zone), or because it runs off towards the bottom of fairly large closed depressions (clearing zone) giving rise to a few temporary pools. However, a fossil hydrographic network has been observed, organised around two main paleo valleys crossing the plateau from east to west: the *Itomboina* corridor (with which the *Ankazomanga* depression is associated) and the *Maroala* valley, which extends as far as the *Ilempo* basin.

The "*Sakasaka*" (Thalwegs) of the coastal strip

The coastal strip between the *Onilahy* and *Linta* rivers has no permanent surface run-off. However, during heavy rainfall, small torrents of water and sand may form. When sufficiently large, these temporary flows can carve out small thalwegs in the sand and sandstone of the coastal strip, creating drainage channels.

The most important "*sakasaka*" and the only ones that can flow all the way to the sea are located in the southern half of the coastal strip, where they have created a temporary hydrographic network.

Tsimanampesots Lake e

In the central part of the coastal strip, along the cliff of the plateau, there is a vast body of salt water, stretching from north to south, over 15 km long and a few km wide. Its elongated shape, along the cliff face of the plateau, is probably due to the fact that it occupies the internal depression at the foot of the Eocene slope, behind an ancient dune belt running parallel to the cliff face.

1.1.1.6.3 The sea

The Mozambique Channel forms the western boundary of our entire study area, covering more than 200 km of coastline, from the mouth of the *Onilahy* in the north to that of the *Menarandra* in the south.

The sea forms a base level for the coastal water table. The influence of the sea can therefore be expected to be very significant, particularly in terms of the hydrodynamics of the coastal zone.

1.1.1.7 Composition of the terroir

As we pointed out earlier, the *Mahafale* landscape is made up of 13 Communes. Six are in the District of *Betioky*, six in the District of *Ampanihy* and one in the District of *Toliara* II.

- The six Communes in the District of *Betioky* are as follows: *Betioky* sud, *Maroarivo*, *Tameantsoa*, *Beantake*, *Masiaboay* and *Ankazomanga*
- The six Communes in the District of *Ampanihy* are as follows: *Itampolo*, *Androka*, *Ankiliabo*, *Ampanihy*, *Ejeda* and *Beahitse*,

- The Commune included in the District of *Toliara* II is that of *Beheloke.*

These thirteen Communes comprise a total of 285 *Fokontany*. Together, they cover a total area of 17,442 km², which represents 26% of the total area of the *Atsimo Andrefana* region.

1.1.2 Population history [Fauroux E., 1989]

Tradition has it that in ancient times, perhaps towards the end of the fifteenth century, a group of zebu herders living in the south-east of Madagascar (probably in the *Anosy* Mountains region) undertook a vast expansion movement towards the immense areas stretching across the south and south-west of the island. Under conditions that have yet to be fully elucidated, this group had benefited from the addition of a complex and coherent body of "magical" knowledge, divinatory techniques and conceptions of political and social organisation, most probably inherited from a core of Islamists who had settled in Madagascar shortly before. This ideological capital, slowly adapted to changing circumstances, seems to have given the group a decisive advantage over the indigenous communities, which were few in number and poorly organised at the time.

The south, south-west and coast were barely populated at the time, and the conquest was apparently not difficult. If tradition is to be believed, the problems arose more from internal divisions within the group, linked to the sharing of wealth and power. These circumstances encouraged early swarming. Each of the swarming groups took with it part of the common capital: zebus, dependents and an ideological heritage that would serve as the matrix for a very particular type of social, religious, magical and political organisation.

The basis of this ideological system was formed by a close and harmonious relationship between a "cult of ancestors", which was undoubtedly very ancient, and dynastic cults which established monarchic power on solid religious and magical foundations. The chiefs of the main Essaim groups were thus at the origin of several dynasties that reigned for several centuries in the south, south-west and west of Madagascar. The political units formed around these dynasties were sometimes at the origin of powerful expansion movements. They were often able to integrate the indigenous populations without too much difficulty, followed by groups of diverse origins attracted by the ability of these units to plunder their neighbours effectively and accumulate wealth in zebus.

In fact, these units, which were essentially political in nature, gradually came to form what was later referred to by the rather inappropriate term "ethnic groups". In Madagascar's southern regional grouping (which extends roughly from Faux-Cap, at the extreme southern tip of the island, to the *Manambolo* river), the *Antandroy*, the *Mahafale*, the *Masikoro*, the *Bara* and the *Sakalava-Masikoro* are found.[18]

The region has always been kept on the fringes, firstly because it offers few opportunities for major development projects, and secondly because its people have had difficulty integrating into commercial relations. Independence seems to have confirmed this marginalisation.

---

[18] Fauroux E. 1989. Une étude pluridisciplinaire des sociétés pastorales de l'ensemble méridional de Madagascar. ORSTOM. Cah. Sci. Hum. 25 (4) 1989: 489-497.

Faced with emergency situations, the Malagasy state has tended to leave these pastoral populations, who are often not very docile and who sometimes support opposition parties, to fend for themselves. As a result, with very few exceptions, the South and West of the country have not benefited from any major development operations, and have not even benefited from the most basic infrastructure investments: the state of the roads, in particular, is particularly lamentable.

1.1.2.1 Presentation of the population

1.1.2.1.1 Geographical breakdown of the population

Around 219,042 people live in the 285 *Fokontany* in the 13 communes that make up the *Mahafale* landscape from north to south. The majority of the population is *Mahafale* (over 80%), followed by *Vezo* and *Tanalàna* on the coast and *Antanosy* (less than 10%) (immigrants who live in certain communes on the banks of the *Onilahy*). The majority of the population in the Cap Sainte Marie area is *Antandroy.*

The following data is available for the 13 communes in our study area. The average population density is 14 inhabitants per km² (national average = 36 inhabitants per km²).

Table 2: Breakdown of the population by component commune of the *Mahafale* landscape.

Sub-region	*Mahafale* landscape	Population (according to PCD)	Surface area km²	Density inhab. /km²	Number of *fokontany*
District	*Betioky*				
CR	*Betioky Sud*	21.763	312	43	26
CR	*Ankazomanga*	22.103	1.647	2	10
CR	*Maroarivo*	22.103	1.658	5	15
CR	*Beantake*	7.142	320	30	12
CR	*Masiaboay*	7.356	485	28	21
CR	*Tameantsoa*	3.709	395	22	13
Total		84.176	4.817	12	97
District	*Ampanihy*				
CR	*Ampanihy*	8.125	1.397	17	25
CR	*Ankiliabo*	12.526	1.134	10	11
CR	*Ejeda*	12.146	1.830	26	49
CR	*Beahitse*	21.190	1.349	15	26
CR	*Androka*	34.683	1.608	16	24
CR	*Itampolo*	41.753	2.417	17	40
Total		130.423	9.735	18	175
District	*Toliary 2*				
CR	*Beheloke*	4.443	2.890	5	13
Total		4.443	2.890	5	13
		219.042	17.442	14	285

Source: Monographie Régionale du Sud-Ouest, 2008.

Poverty has affected a larger proportion of the population in recent years. Indeed, the poverty ratio[19] in the *Atsimo Andrefana* Region rose from 75.2% in 2009 to 82.1% in 2010 [Cellule

[19] Or the incidence of poverty, which measures the percentage of the population living below the poverty line .

Régionale de Centralisation et d'analyse, 2010]. The human development index[20] for the area was estimated at 0.399 in 2005 (0.527 at national level). All this illustrates the precariousness of living conditions in this region. The low level of education is one of the most important factors holding the region back. Less than 15% of heads of household have attended secondary school, and more than 57% have never been to school. Infrastructure and basic living conditions are very limited: access to water and energy, health, education and vocational training, communications, roads, access to credit, etc. Over 80% of the population drink water from unprotected traditional wells, unprotected springs and rivers. Women and children are sometimes forced to make long journeys on foot or by cart to obtain water.

1.1.2.1.2 Migration

Our study area has a high level of migration. People (students, workers, farmers, etc.) come to *Toliara* or *Morondava* for various reasons.

Emigration mainly concerns the *Mahafale, Tanalana* and (to a lesser extent) the *Vezo*, who make up the bulk of the population. Migration phenomena can be classified according to their duration: seasonal migration lasting no more than a few months, temporary migration when the absence lasts only a few years, and permanent migration when the emigrant has settled permanently outside the country of origin.

They can also be classified according to distance:

- ✓ short-distance migrations (as far as *Toliara* or *Ampanihy*),
- ✓ medium-distance migration, to the *Bas-Mangoky* region of *Morondava* and in the Sapphire Lands,
- ✓ Long-distance migrations, for example, to the sugar cane plantations of *Nosy Be* and the *Mahavavy* in the north. *Belo* sur Mer and *Maintirano* for the *Vezo.*

The nature and direction of migratory phenomena vary from one part of the coastal plain to another.

1.1.2.1.2.1 Seasonal emigration (short-distance migration)

This is by far the most important type of emigration in terms of the number of people involved each year.

In the rainy season, you have to be on the spot to plant and look after the herds during the transhumance. But as soon as the dry season arrives, once the harvest on the *tetikala or tonda* (fields) has been completed and the herds have returned from the pastures in the east, there's not much left to do in the village. Every year, a large number of young people between the ages of 15 and 20 leave the village. If the harvest has been poor, older people also leave in large numbers. Those with a lot of oxen feel less need to emigrate. Seasonal emigration therefore appears to be a phenomenon that recurs every year in the dry season, but to a greater or lesser extent, depending on the climatic conditions that have prevailed on the coastal plain in the preceding months.

1.1.2.1.2.2 Temporary and permanent emigration

[20] HDI: the ability of the area's population to live long and healthy lives, to be educated and to enjoy decent living conditions.

It is often difficult to distinguish between temporary and permanent emigration. Emigration to the north of the island is always with the intention of returning.

1.1.3 Socio-political organisation

The population of this region is ethnically homogeneous; in this respect, we are witnessing a similarity in terms of the organisation and functioning of society.

Social organisation is essentially based on kinship (blood ties and marriage) and age group (seniority).

It is a lineage-based society, at the base of which is the lineage *(Tarika),* followed by the clan[21] *(Raza)* and ending at the top with the *"foko".*

The clan is, however, the essential element. Property is therefore clan-based. At one time, property par excellence was the herd. Each clan had its own herd and its own ear tags for the zebus, the *"vilo".* On a political and religious level, the clan also had its own specific oral prescriptions, the *"lilindraza".*

The clan chief, sometimes called the *"mpitata"* or *"mpitankazomanga",* was a political leader, supreme judge and, above all, a religious leader. He was the guardian of the *"Hazomanga"* tutelary poles, at the foot of which ritual zebu sacrifices, known *as "soro",* were performed.

This is another reason why the zebu is so important, because historically, when clans emigrated from the East and based their power on the size of their herd, they were able to cope with the demands of the ceremonial cycle.

1.1.3.4.1 Kinship

The people of this region live in village communities in hamlets where families from the same lineage are generally grouped together (this situation also applies in the outlying *Fokontany*). The community is represented by the eldest[22] , a direct descendant of the founding families, following the paternal line.

Tradition and family property are passed on by men who marry women from the same ethnic group, but from a different lineage.

1.1.3.4.2 Age group

If we look at it hastily, this age group appears to be influenced by the Western structure. Nevertheless, power is gerontocratic and the traditional hierarchical organisation has always been respected as far as each family is concerned. The status assigned to each individual is determined by their respective age.

"The Mpitan-kazomanga

Represented by the eldest descendant of the founding family, as his name suggests, the *Mpitan-kazomanga or Mpitata* is the guarantor and keeper of ancestral tradition. He plays the role of mediator between the ancestors and the living, which gives him decision-making power. Consequently, in the social production process at this level, the community is not directly

---

[21] In Malagasy anthropology, the Raza can be defined as a historical community of descendants.

[22] Mpitata or Mpitan-kazomanga

involved in the production of material goods, because the means of production are privately owned, hence the independence of each *"Tokantrano"*.

The *Mpitan-kazomanga* is also the guarantor of the resulting social cohesion. Social cohesion is also ensured by customary rules that are revived every two years by a social pact materialised with the help of *Soro*[23] . All the men in the community must drink the water from the gold and eat the goat meat while listening to the prayers of the village elder. The main customary rules in force are :

- The social pact *(titike),*[24]
- Fire and cooking pots must not be taken out of the hut at night,
- A ban on night-time pounding,
- A ban on harvesting plants marked with *vorovoro.*[25]

If these rules are disobeyed, misfortune befalls the family concerned. To remedy this, zebus or chickens must be sacrificed. These rites are called *Hifikifike*[26] .

There are sacred places around each village, the most famous of which are *Ankilifaly*[27] and *Anjampaly*. People say that spirits called *Tambahoake*[28] inhabit these places. These spirits may inhabit a person, a phenomenon known as *KOKO*. The person concerned is taken out of the village by the spirits for a period of 3 months to 1 year. On their return, they must undergo a ritual called *Vakiendela*[29] or *Aboake*[30] , which lasts three days and in which music plays a major role. The purpose of this ritual is to give the person concerned an opportunity to recount what they have experienced and to hear messages from *Tambahoake*.

People go to sacred places to communicate with the spirits through various rites, in order to obtain a cure for an illness or good fortune. In the case of healing rites, the sick may present themselves directly to the spirits or through the possessed person. They bring a black chicken or rooster that has been marked, i.e. from which the hair[31] around the throat has been removed before it is set free. The population must also respect the following prohibitions:

- ✓ Do not carry axes or sandals near sacred places,
- ✓ Do not add salt to the milk.

It has been observed that sacramental rites (*e.g. tromba*[32]) and ceremonies (e.g. funerals) are held mainly from July to October.

---

[23] A rite to strengthen cohesion among members of the community, marked by the tasting of water soaked in gold with the meat of a goat.
[24] *Titike* = ceremony designed to make a collective agreement more solemn, but concerning a geographical area that is generally more restricted than the *dina*.
[25] Dried grass or pieces of linen tied to a schooner planted in the fields to tell communities that this is private property (to mark a traditional self-born).
[26] A rite that puts the blood of a chicken or zebu on a person's forehead to purify them of their misfortune. The mpitan-kazomanga ensures that the person in question is purified.
[27]Located to the north of Miarintsoa. Important trees in the area: A dozen tamarind trees
[28] Known by Christians as Lucifer.
[29] Loosen your tongue.
[30] Presented by the soothsayer before the Community.
31 This means that the rooster represents a person who is ill and that their illness is removed by Tambahoake and they are completely cured, so they can resume their daily tasks.
[32] Tromba = the person possessed by a spirit from the sea

1.1.3.4.3 Conflict management

There are two ways of solving individual or collective problems in the *Mahafale* community.

The first possibility is to take the matter to the *Fokontany* chief, so that he can mediate. If there is no reconciliation, the *Fokontany* president refers the case to the *Hazomanga* elders. If they are unable to bring the conflict under control, they must refer the case to the *Fokontany* president, who convenes the *Fokonolona* (General Assembly of the village population), which will assist the court under the tamarind tree known as *the "Tsara"*. The parties in dispute must then pay the sum of 20,000 *Ariary* as a fee, 10% of which goes to the *Fokontany* president and the rest is divided between the existing lineage groups. If no solution is found, the *Fokonolona* demands the sum of 60,000 *Ariary* to draw up a Procès-Verbal of non-reconciliation.

The second option is to take the matter directly to the traditional authorities. If these people are unable to settle the dispute, they can call a meeting of the *Fokonolona* directly, without having to go through the president of the *Fokontany*. The amount to be paid varies according to the appreciation of the elders, and the part that goes to the elders is called *"Lohalily"*.

1.1.3.4.4 Clan divisions

The *Mahafale* are divided into a large number of clans, a list that may seem tedious. However, it is essential to describe this traditional clan organisation in order to understand certain geographical facts: for example, the seasonal movements of the herds, which are organised according to these clan divisions.

The clan, as an essential element of the social structure, has many geographical implications.

1.1.3.4.5 Regional and local socio-economic characteristics influencing the forest resource conservation project:

An anthropological approach reveals a certain complexity that needs to be clearly identified in order to develop a more appropriate strategy for future decision-making and management of protected areas (park and TGRN).

An initial approach enabled us to identify subdivisions within the two major anthropological groupings, *Mahafale* on the plateau and *Tanalana* on the coast, and to identify the decision-making centres, the major dynasties and clans that dominate and control major decisions, the clans that are dependent on the dominant clans and dynasties, the recalcitrant clans that do not submit to the domination of the dominant clans, the minor clans, the marginalised groups, the destitute, and so on.

The various decision-making blocks, which generally correspond to the dominant clans whose sites and holders of the *"hazomanga*[33] " or royal sites and *"mpanjaka"* kings, *the* centres of major decisions, need to be clearly identified.

A more in-depth approach enabled us to go into more detail.

1.1.3.4.2 Remote occupation of space and the traditional decision-making system :

[33] Ritual pole at the foot of which all the clan's traditional ceremonies take place, including traditional marriages or "fanambalia", offerings to the ancestors through the sacrifice of zebus or "soro", etc. under the office of the *"mpitan-kazomanga"* who is the holder of the ritual pole, playing the role of "priest" in a way, who is also the master of the clan's major decisions.

- Among the *Mahafale* people living near protected areas (park and TGRN):

The *Mahafale* living space was originally organised along the river *Linta,* then along the axis of the national road 10. Pastoral areas were not yet a problem. There were several reasons for the expansion of the *Mahafale* plateau:

While in *Tanalana* territory the clan segmentations are clearer and easier to locate, and the notion of disadvantaged groups and recalcitrant clans less obvious, in *Mahafale* territory the situation is much more complex and the notion of dominant groups, dependent groups, marginalised groups and deprived groups is very obvious. We will try to describe the hierarchy of power in *Mahafale* society in the "kingdom of *Onilahy*"[34] .

North of the River *Linta*, the hypothesis that the *Mahafale* area north of the River *Linta* was occupied from south to north[35] seems to be confirmed. In fact, living areas are mainly located in the southern part of this area, which forms the municipality of *Ejeda.*

The central part of this area is still sparsely inhabited and is used as grazing land by the surrounding herders around the *Ilovo* depression pond to the west of *Beahitse* and the *Ilembe* pond much further north of the *Ilovo* depression.

During the first half of the 18th century[ème] , the *Maroseragna* kingdom of *Menarandra* was at its height.

King *Tsimamandy*, also known as *Lahimanjaka*, considered a great conqueror and organiser, led territorial conquests towards *Onilahy* and was at the origin of the "*Maroseragna* kingdom of *Onilahy*". This vast area was occupied in a scattered manner by groups considered "*tompon-tane*", the first occupants, such as the *Tenagnaniha* of *Bekinagna*, the *Tandenta*, the *Renelimy* made up of the *Andrianaivo*, the *Andriambato*, the *Talaotse* also known as the *Teaby*, the *Andriambetsiaohatse* and the *Andriatsileleke*.

Following a compromise with the *Renelimy*, through a matrimonial strategy, *Tsimamandy* placed his two sons at the origin of the *Maroseragna* kingdom of *Onilahy*. They took up residence at *Sakoatovo*, a site with all the advantages of agro-pastoralism: proximity to water, good alluvial land and beautiful forest pastures.

At present, the royal *Maroseragna* clan of the "kingdom of *Sakoatovo*", which has become enlarged, has split into two, with one segment still living in *Sakoatovo* and the other segment in *Ankazotà*. No conflicts were reported as a result of this segmentation.

Alongside the *Maroseragna* kings, further south, the *Manindrarivo* kings of *Ankiliabo* and the powerful *Faloagnombe* blacksmiths of *Manakaralahy* and *Manakaravavy*, suppliers of weapons to the *Maroseragna* kings, pushed back the small groups that had joined them northwards to the high plateau of *Ampitanake.*

- Local specificity around *Beahitse :*

---

[34] The Maroseragna kingdom of Onilahy, whose royal site is Sakoatovo, is geographically located between the river Manakaravavy, 22 km south of the town of Ejeda, which forms the boundary of the municipality of Ejeda to the south, and the river Onilahy to the north. In relation to the "Maroseragna kingdom of Firangà" to the south of the Manakaravavy river

[35] From Zafiraminia

The commune of *Beahitse* was created out of the need for independence on the part of livestock farmers. Three clans are most well known: the *Telavaposa* from the almost abandoned village of *Lavaposa* near *Itampolo*, the *Tevato* and the *Zanakanga*. These are cattle-breeding clans who have acquired social notoriety over the years with the prosperity of their cattle farming. These three dominant clans still take turns in the official administration of the commune.

- The *Tanalana* confederation "***Tokobey telo***[36] " on the coast:

According to oral sources, towards the end of the 15th century[ème] and during the first half of the 16th century[ème] , clan and lineage traditions refer to the migration of small groups of people from the south-east (Fort-dauphin) to the south-west. The reasons given for this include internal conflicts and the search for new areas. The history of the settlement of the *Tanalana* coast can be traced back to two brothers, *Etsivale* and *Renioma*, the founding patriarchs of the *Tetsivalea-Temitongoa* and *Tevondrone* clans, who were part of these cohorts.

Enioma, a diviner and healer, continued his ascent northwards and settled with the *Tetembola* of *Anantsono* (Saint Augustin). His son *Earere,* in disagreement with his maternal uncles, set off south again and settled in Vondrone (a favourite place for *vondro/* rushes) in the vicinity of *Elosy*, occupied by the *Tetsivalea*. The clan's name derives from the space: *Tevondrone.*

Earere's descendants, from his first wife, formed the *Tevondrone* clan, divided into several lineages:

- *Emihela,* the eldest, settled near Lake *Tsimanapetsotse*, where a lot of '*vondro*' [Cyperus, rush] grew, probably the origin of the name '*Tevondrone*'; he headed the *Berohala* lineage there;
- *Ehilala*, *Earere*'s second son from his first wife, based in *Maromitilike*, led the *Tantohatse*;
- *Emiheritse*, the third son, in *Ejerea* (*Marofijery*) administered the *Tetsialahatse* ;
- *Emitoha*, the last son, north of *Soalary* was the patriarch of the *Tekazonosy*.

The presence of a *Temahaleotse* enclave in this area can be explained by the fact that during his lifetime, *Emihela* settled his children from his second wife and those of his sister, married to an *Antandroy Karimbola*, in an unoccupied area between *Montilimy* and *Vatolalake* in order to avoid succession problems. We will describe below the fate of *Earere*'s descendants from his second wife, the *Tanalanapote.*

Etsivale, authorised by the *Tevatoaore* led by the patriarch *Andriamianto* and considered a "*tompon-tane*" (landowner), settled in *Elosy*. Today, its descendants, subdivided into two middle clans, are the *Tetsivalea*, who occupy the villages of *Ankilibory-Avaratse*, *Antanendranto* and *Ankilitelo* respectively, and the *Temitongoa*, who occupy the villages of *Besely-Nord* and *Nisoa* up to the present-day village of *Andoharano*. Certain lineage groups such as the *Tesendre*, the *Temafaitse* and the *Tomiriafara*, considered to be *Mahafale*, live side by side in this vast territory.

As for the *Temilahehe*, at the same time, towards the end of the XVème and beginning of the XVIème century, it seems that an individual from the village of *Amparehetse*, about 5 kilometres west of *Ambovombe*, crossed the *Manambovo* and sought refuge with the *Antemanatsà* of

[36] Literally "Tripod" to designate the "three great hearths: Tevondrone, Temitongoa, Temilahehe".

Ampalaza, who gave him a wife and nicknamed him *"Tsimalilo"* [who never gives up]. His two descendants, *Lailava* and *Ehosy*, left *Ampalaza* and settled in the *Elosy* area, abandoned by the *Tetsivalea*.

The *Temilahehe,* descendants of *Ehosy* - whose posthumous name is *Andriambeloza* - now occupy the south of *Nisoa: Kaikarivo*, where their "*hazomangalava*" is established, *Befengoke, Ampitanake, Befolotse, Andoharano, Androimpano, Tsiamena, Ankamenà, Betratratra-Besono, Tsiandriona* sud, *Ankiliambany, Ambaladoda, Belambo* On the left bank of the *Linta*, lineage segments have settled in *Saodona, Mandevy, Tsiarindrano, Zàmasy, Beharahake, Manera* and *Tanamilitsy.*

Not only are the *Tetsivalea*, the *Temitongoa* and the *Tevondrone,* of *Antanosy* origin through their paternal ancestry, the *"ziva"* (joking relatives) of the *Antandroy*, but on the coastal plain a matrimonial strategy is in operation whose instigators are the respective patriarchs of the three major clans involved: *Andriamaharavo*, for the *Tetsivalea-Temitongoa; Andriamiela*, for the *Tevondrone; Andriambeloza* for the *Temilahehe.*

A pact of non-aggression and mutual assistance finds its raison d'être in the establishment of the "*Tokobey telo*" clan confederation, the tripod which tacitly governed and still governs all the population groups of the *Alanana*, a sandy coastal plain. *Tanalana* means "people of the *Alanana* or *fasika*".

- Two groups of lesser importance: the *Tandroka* and the *Vezo sarà* :

The *Tandroka*, who group together, by affinity, all the natives of *Androka*, including the *Vezo,* the *Temilahehe*, and the *Tevondrone* natives of the *Tevondrone* maritime coast. Here, we were told that the importance of the clan has very little influence on major regional decisions.

The *Vezo sarà* are an allochthonous group of seasoned fishermen who migrated from the north to the south as far as *Androka*, setting up fishing camps in places where fish are plentiful, camps that have now become villages.

1.1.3.4.3 The recent conquests of new areas :

- *Tetsivalea,* cramped for space:

The *Tetsivalea* are great cattle breeders, but they are somewhat cramped in relation to the edge of the cliff, which was originally a Tertiary fault. Although there were no historical conflicts between the *Tetsivalea* and the *Tevatoaore*, certain internal lineage disagreements within the early *Tetsivalea* created a segmentation of the original clan, giving rise to three clans: *Temanambone, Tetsivalea* and *Temitongoa*, each representing a decision-making centre.

These three clans generally lived together peacefully. This did not exclude certain conflicts over territorial interests, which were of course of little consequence, but which led to the departure of certain groups to conquer new lands in the forest. This explains, among other things, the creation of the *Nanohofa Fokontany* and its hamlets; it also explains the semi-permanent migration of many *Tetsivalea* to the town of *Toliara.*

By way of clarification, the *Tetsivalea* live in the area between *Tanendranto* and Andranovao (Commune of *Itampolo*). In fact, if we make a classic East-West transect which schematises the territory of the *Tetsivalea* :

The width of this sandstone-sandstone coastal strip varies from a few kilometres at *Andranovao* to around fifteen kilometres at *Ankilitelo*. From west to east, we find :

The dune shore is little used by the *Tanalana*. On the other hand, it is the place where the *Vezo* fishermen settle, and we mentioned above the latent conflict that persists between them and the *Tanalana.*

When flooded in the rainy season, the salt depressions are used as watering places for zebus.

The very dry "*tane maiky*" soils are humus-free soils, generally sand-gravel mounds of ancient dunes of Flandrian or Aepyornian age, or sand-gravel-limestone outcrops. These are known as *"monto"*. Villages are generally built on them.

Monto" is generally uncultivated land, covered with a very low cover of grasses, if not mostly thorny shrubs, including Euphorbia oncoclada "*samata*" used as fodder for zebu, combined in places with other thorny shrubs to form xerophytic thickets used as grazing land for small ruminants.

In the area close to the "*olo-bohitse*" cliff, which varies in width and is the widest, around 10 km, at *Beroy* and *Anjà-Belitsake*, from west to east we have in succession :

Humus-rich "*tane jôbo*" soils were created by slash-and-burn clearing of the original thickets that covered a large part of the land. It should be noted that these soils are not rich enough in humus and are often abandoned after a few years of cultivation. The cultivated land is laid out in *"vala"* enclosures against ruminants. Once abandoned, this land becomes *"sarike". Sarike"* can be farmed again if the thin humic layer regenerates, otherwise they become *"monto"* and are abandoned for good.

In the area around *Vohombe* and *Tanendranto*, the *"tane jôbo"* are used to grow cassava for food, along with watermelons and other Cucurbitaceae; in the southern part of the territory, among the *Temilahehe, they are* used to grow cassava, lentils, sweet potatoes and vohème peas, always combined with watermelons and other Cucurbitaceae.

Beyond the *"tane jôbo"* zone, as far as the edge of the cliff, there is an increasingly narrow strip of forest, where new *"tetike"* slash-and-burn clearings are planted for food crops such as cassava and maize, or for cash crops such as tobacco and, no doubt, poppies.

- Forests as places of refuge :

The land belongs to the royal or dominant clan. It has full rights over the land, including the right to bequeath part of it to groups that are "interesting" to it. It also had the right to expel from its lands groups that had "grafted" themselves onto its clan. Minority groups sometimes found themselves driven off their original settlements, moving into the forests without a landowner, as there were many empty forest areas where the traditional owner of the land was the person who first exploited it.

- In *Mahafale* territory:

On the one hand, in its initial structure, the *Mahafale* population was made up of the "*renin-tane*" [first occupants, members of the *Maroseragna* dynasty], the "*valo hazomanga*" [commoners of various origins attached to the royal family and to the person of the king], and the "*folo hazomanga*" [freedmen of various origins, benefiting from royal protection, who settled from *Ampalaza* to the *Antengy* well on the south coast of *Ampanihy*].

Under royalty, the colonial administration and even the First Republic, for all minorities, political dissidents and/or groups fleeing the burden of taxation, the forest plateau became an area of refuge.

On the other hand, for the *Maroseragna* kings, the major decisions were taken by the dominant "*renin-tane*" clans[37] . The *Tampatsy* or *Tampatsiagnombe* are a marginalised group alongside the "*renin-tane*", apparently because the group never succeeded in cattle farming. In fact, the name "*Tampatsiagnombe*" means "curse on cattle rearing", so much so that the *Mahafale*, a traditional cattle breeder, never entered into matrimonial relations with this group.

In addition, the *Manindrarivo* kings of *Ankiliabo* [near *Ampanihy*] and the powerful *Faloagnombe* blacksmiths of the *Manakaralahy* and *Manakaravavy* area, suppliers of weapons to the *Maroseragna* kings, were gaining more and more space and notoriety, This is probably the origin of the *Tezamainty* who created the village of *Vorojà* and who seem to have come from the *Etakake* sector.

However, it should be borne in mind that certain *Mahafale* forests were sacred and could not be used as a refuge. These include the *Ankirikirike* forest, the necropolis of the *Maroseragna* dynasty of *Menarandra*, and the *Manintsy* forest which sheltered the mortal remains of the *Maroseragna* of *Onilahy*.

In the not-too-distant past, around the 1970s, when the Société Nationale des Huileries de *Toliara* (SNHU) was in operation, the popularity of groundnuts was everywhere. This led to a scramble for the ferruginous red tropical soils that were good for groundnuts. People were clearing the land and producing groundnuts. New settlements were formed in the forest towards the interior of the plateau, and over time became hamlets that grew into villages, which in turn became *Fokontany* chief towns.

- In the western sector of the commune of *Beahitse* :

In the XXème century, the *Telavaposa*, *Tevato* and *Zanakanga* became the clans of the great cattle breeders who were masters of the major decisions. To this day, they take turns in the official administration of the commune.

Those who did not belong to these three clans were subject to their decisions and were pushed westwards into the forest *no-man's-land* to the west and north of the *Ilovo* watering hole. There they set up clearing camps, which became the present-day villages of *Beomby*, *Satrapaly* (*Marofototse*), *Ankalirano*, *Belombiry*, *Esifake* and *Farafatse*.

- The case of the *Tanalanapoty* of the commune of *Beheloke* and *Maroarivo*:

Earere's descendants, from his second wife, formed the *Tanalampoty*, who were pushed northwards into the commune of *Maroarivo*, some of whom reached the *Onilahy* valley.

However, one of the branches managed to enter into a preferential marriage with the *Temitongoa* from *Itampolo*, which strengthened their position in *Manasy* in *Tevondrone* territory. The *Tanalanapoty* of *Manasy*, although socially well structured, are unfortunately somewhat on the fringes of the major decisions regarding the development of the *Fokontany*. Moreover, the administration of the *Beheloke* commune is still attributed to the dominant *Tevondrone*.

---

[37] Literally "mother of lands".

- The people of *Ambolisogno* and *Valavo* :

This allochthonous group was originally "grafted" onto the various segments of the *Temitongoa* of *Nisoa*, descendants of *Betsiriry*, for "services" linked to cattle rearing. Moreover, this group was considered "*ana'i Betsiriry*" [children of *Betsiriry*].

The demographic expansion of the *Temitongoa* of *Nisoa* "pushed" the group towards the current forest territory of *Ambolisogno*, located further east, around 6 km from *Nisoa*, and in the middle of the *Valavo* forest. The village of *Valavo*, made up of six hamlets, despite being located 27 km from *Ambolisogno* on the *Ampitanake* track, is still administratively attached to the *Ambolisogno Fokontany*.

At present, the inhabitants of *Ambolisogno* have apparently been forgotten by the *Itampolo* communal administration, which is currently run by the *Temilahehe*. The *Fokontany* has neither a water point nor a school. The nearest water point is at *Nisoa*, where the people of *Ambolisogno* and *Valavo* draw water; and only three people in the villages of *Ambolisogno* and *Valavo* are literate.

1.1.3.5 Socio-cultural components influencing the conservation project :

1.1.3.5.1 Habits and customs :

- The habits and customs of the *Mahafale* :

Here, the zebu-centred civilisation is more than visible. In *Mahafale* country, the zebu is a quasi-sacred animal:

- the zebu pen is always on the north-east side, the direction taken by the spirit of the deceased at the moment of death, the side of the ancestors, and it is forbidden, even taboo, to 'defile' this part of the village;
- the use of zebu dung as manure is forbidden, even taboo; moreover, piling it up at the zebu pen site is a sign of blessing and social notoriety;
- It is taboo to put salt in cow's milk, as this would decimate the herd; moreover, it is forbidden to drink milk standing up or to drink milk with one's head covered or feet shod.

For the *Mahafale* man, the ultimate capital is the zebu, and other income-generating activities are merely means to this end. Zebu capital is a sign of blessing, and therefore of moral satisfaction and well-being, social notoriety and wealth. Some clans have been able to rise to a higher social hierarchy thanks to zebus. Moreover, the "*mpagnarivo*", the owner of a few hundred zebus, is a very influential man in his society.

The land tortoise is taboo everywhere for the *Mahafale*.

- The habits and customs of the *Tanalana* :

There are similarities between *Tanalana* and *Mahafale* habits and customs in terms of the "*hazomanga*" structure and traditional rituals.

Tanalana society is a clan and lineage community. The social structures that make up the "*Tanalana* confederation" will be described later. We can already say that this confederation is made up of three major clans: the *Tevondrone,* the *Temitongoa* and the *Temilahehe.*

In terms of social organisation, the descendants of each of the three major clans - *Tevondrone, Temitongoa, Temilahehe* - were socially grouped around a "*hazomanga lava*" or "*hazomanga lahy*" [central ritual pole] which was both a sacred symbol and a traditional "place of worship" for offerings to the creator God *Zanahare* and the ancestors *Raza* whom the *Tanalana* venerate, and where the service for the offerings was the responsibility of the "*mpitan-kazomanga*" [the holder of the ritual pole], an exclusively male responsibility that is passed down from generation to generation to the first son of the eldest in each generation and is therefore not necessarily the responsibility of the eldest. However, the oldest members of the clan, the '*olobe*', are the 'wisdom' of the group.

From the three major clans mentioned above, with demographic growth and the distances separating families, "*fokoe*" [lineages] have been created, some of which have organised themselves into *"hazomanga fohy"* [secondary ritual post].

As far as religion is concerned, the vast majority of *Tanalana* people practise traditional religion based on ancestor worship. This cult attributes divine powers to the ancestors. Like all Malagasy, the *Tanalana* respect the dead. They believe that the *"avelo"* spirits of the dead never die and watch over the living. Conversing about family events with the spirits of the dead is therefore a way of asking for a blessing or favour.

Animals are sacrificed and offered to the ancestors at the *hazomanga* site on the occasion of family events, or to ask for blessings, favours, prosperity or, quite simply, health. The zebu is the animal of choice for offerings, while sheep or roosters can be used for minor events.

The main family events, such as the "*soro*" zebu sacrifice [zebu sacrifice to ask the *Raza* for blessings, to thank them for favours obtained], the "*savatse*" circumcision, the "*fandeo*" marriage proposal, etc., must take place at the "*hazomanga*" site according to well-defined rituals orchestrated by the "*mpitan-kazomanga*", holder of the ritual pole.

Malagasy civilisation, and that of the *Tanalana* is no exception, is centred on the zebu. Raising zebus is one of the *Tanalana*'s main activities. As elsewhere in Madagascar, the animal is used to "please" the ancestors.

The zebu is also the draught animal for the cart, a local means of travel and transport.

The *Tanalana* would never agree to be buried without a hardwood coffin, the "*mendoravy*" [*Mendoravia tulearensis*], otherwise the spirit of the deceased would create misfortune and death among the living.

And let's not forget that the "*kotroke*" land tortoise [*Geochelone radiata*] is taboo everywhere, even to the touch.

Foreigners must respect this taboo, as the *Tanalana* would never tolerate any breach of it. A certain reticence towards foreigners, for whom the tortoise is not taboo, is sometimes observed for this reason.

The strategy for protecting turtles should undoubtedly be reviewed, because two interests are at odds with each other:

- On the one hand, local people want their territories to be "emptied" of undesirable turtles;

- conservationists who are working hard to save the species.

We would like to highlight two rituals, one of which, the "*tsotse*[38] ", is directly linked to climate change and the other, the "*soro an-kazomanga lava*[39] ", to strengthening the powers of the major decision-making clans.

In conclusion, the clans on the *Mahafale* plateau vary greatly in size. Whatever its size, each clan has a large *hazomanga*, the clan altar where the *Mpisoro*, the clan priest, resides. This altar is made up of a series of poles, generally made of *katrafay* wood, planted vertically in the ground and around 2 m high. One of these poles has a horizontal board attached to its pointed end, on which pieces of the liver, tenderloin or hump of the consecrated zebu are displayed after the sacrifices.

The clan is a living reality and, consequently, a changing one, under the pressure of geographical causes, but also, more frequently it seems, as a result of personal or lineage rivalries.

1.1.3.5.2 Funerals

Funerals are of great cultural importance in *Mahafale* society. They are organised in three phases.

Photo 1: Stone covering work on the tomb after burial

1.1.3.5.2.1 First phase: burial *(ahaja)*

The immediate family of the deceased must sacrifice zebus and goats[40] . The number of these animals varies according to the economic

[38] *Tsotse*" is a special "*soro*", an annual sacrifice of the best zebu to ask for rain.

[39] Also known as "*sorombe*": Periodically, the members of each major clan - *Tevondrone, Temitongoa, Temilahehe* - gather at the site of the "*hazomanga lava*" or "*hazomanga lahy*" for a great sacrifice, an offering of "*sorombe*" zebus to the god *Zanahare* and the ancestors *Raza*. This is an opportunity for lineages and members of the extended family to get to know each other, thus confirming and strengthening social cohesion. It is also an opportunity to prove and/or strengthen the clan's power.

[40] The family of the deceased kills oxen and goats to give people who are going to attend the funeral and people from far away the opportunity to express their condolences to the family of the deceased. The slaughtered oxen and goats symbolise the flesh of the deceased for the community.

power of the officiants. The cutting of *Mendoravy* (*Albizia tuleriensis*), A tree used to make coffins *(Hazondolo)* and build tombs, requires the sacrifice of castrated zebus.

1.1.3.5.2.1 Second phase: preparing the funeral ceremonies

The family of the deceased is looking for money to finance the ceremony. They invite close family and friends to attend the ceremonies and contribute to the funeral expenses.

1.1.3.5.2.2 Third phase: funeral ceremony - (*fihisà*)

This is a time of festivities. All the guests bring goods to the families close to the deceased. Specifically, sons-in-law and sons-in-law must offer at least one zebu. As well as livestock, other goods may be offered, such as chairs, beds, suitcases, bicycles or sewing machines. This gesture is called "*enga*[41] " and is accompanied by dance.

In return for these gifts, the family of the deceased must offer goats or food to the guests. This gesture is called *"famaha"* or *"Laobary"*. As the saying goes, the reward will depend on your contribution *(Arakarake ty enganao ty mahasoa ty laobarenao).*

Fihisà" lasts three days. The *"Fanengana*[42] *"* and *"Famaha"* take place on days 1 and 2ème . On the 3rd day, two zebus are taken to the burial site where the deceased will be given a final burial. The zebus are sacrificed to be eaten by the guests. Zebus donated by relatives may also be killed, if the family of the deceased so wishes. The horns are used to decorate the tomb and are a sign of social rank.

After the ceremony, the family burns or destroys the house of the deceased, and changes the name of the deceased according to his behaviour in life, his age, and his sex, adding the following suffixes:

- ✓ Young man : *Laza*
- ✓ Girl/Woman: Vola
- ✓ Old man : *Arivo*
- ✓ A person who liked to fight: *Aly*

For example: *Alivola*: a girl or woman who likes to fight.

1.1.3.6 Village territory

Each village has a territory called a *"faritany"*, the boundaries of which, marked by trees, grass, stones or other easily identifiable physical objects, are well known to the *"fokonolona"* (communities).

---

[41] The act of bringing things for one of the members of the deceased's family. The thing brought is either ordered by the person (something they wish to have: for example a goat or a castrated billy goat; a 1 year old zebu or many others) or material goods such as plastic chairs, a sewing machine, mattresses etc. In general, material goods are ordered or intended for women and oxen and goats are intended for men. If a person has received an ox or a goat, it must be accompanied by a sum of money because this sum has a particular meaning: *"tadiny"* is for tying up the zebu so that it does not escape. This person will receive a gift in return from the owner of Enga.
A person who has brought his "Enga" will wait his turn because in his family, in the event of death, he must expect to receive a 2-year-old zebu if he has given, for example, a 1-year-old zebu. If the person who has received a 1-year-old zebu is unable to bring his gift (this is known as *"Mamaly"*), their relationship immediately deteriorates.
[42] Carrying out its Enga

The *"faritany"* is experienced by the villagers not as a neutral geographical support, but as a space with which they have a mystical as well as agro-economic link.

It is a space where the sacred is strongly present, because it is in this sociological space that the dwellings of the deceased, the tombs, the *"hazomanga"* sacrificial poles, *the* circumcision poles, the places of worship, but also the places of meetings and other manifestations of community sociality accumulate.

Photo 2: *Mahafale* tomb decorated with *Aloalo*

The *"faritany"* is also an economic area that corresponds to developed or developable land, but also uncultivated areas of little agricultural value, including grazing areas and farmland. The grazing areas are divided only by territorial boundaries that correspond to those of the village, or *"faritany"*. For the coastal peoples, the areas where they often go fishing are part of their *"faritany"*. Each clan has been able to identify the boundaries of their marine territories.

Land appropriation only concerns agricultural land and is carried out by lineage. Possession often only means right of use at the level of the lineage fragment or the small family. Every two or three generations, there may be a redistribution between brothers, heads of small families [RAZAKANDRENY C, 2005].

1.1.3.7 Challenging traditional clans

The foundations of the traditional system, based on the clan family and ancestor worship, are now being called into question. The clans have become too large due to demographic pressure, their former religious centre is tending to fall into disuse and there are fewer and fewer candidates for the position of *"mpitan-kazomanga"* or *"mpitata"*.

A social divide is developing between the *Mahafale* who continue to live in the villages and those who have left to seek new opportunities elsewhere.

However, the *Tanalana* clans have been able to resolve this divide by forming a confederation called *"Toko Bey Telo"*. Periodically, the members of each major *Tanalana* clan (*Temilahehe, Tevondrone and Temitongoa*) organise a *"sorom-be"* or *"soro an-kazomanga lava"*. They gather at the foot of the *"hazomanga lava"* or *"hazomanga lahy"* for a great sacrifice, offering zebus to the Creator God (*Zagnahary*) and to the ancestors (*Raza*). This is an opportunity to renew *the* pact of non-aggression and mutual assistance to strengthen social cohesion.

1.1.3.8 Religion

There are religious buildings in our study area. These buildings are Catholic and Protestant (the *LWF*). As far as religion is concerned, apart from the ECAR and the *FLM,* the communes also have the Adventist church, the Pentecostal church, the Assembly of God church, the FJKM and the *Jesosy Mamonjy*. The Muslim religion (Shiite) is beginning to make inroads in the district capitals, particularly in *Betioky*.

As for daily practice, there are customs and habits such as circumcision; people also worship *Tambahoaka* or *Bilo* or "*fanompoa tromba*".

1.1.4 Economic activities

1.1.4.1 Operating mode

Firstly, the farming method is direct farming. These farms are small, ranging in size from 0.5 to 5 hectares, with the majority being around 2 hectares, although the available land is not fully exploited due to a lack of resources or initiative, or for reasons of land ownership.

In addition, the contribution of family members is limited to certain types of operation, hence the use of external labour: salaried workers.

Salaried work: initially, it was the members of landless families affected by the food shortage who provided salaried agricultural work. The employee is employed as a day labourer, with a verbal contract, carrying out a very specific task. According to the contract, payment for the work done is made in cash at the rate of 3,000 *Ariary* per day.

In the *Mahafale* landscape, the main economic activities are agriculture, livestock farming, commerce, handicrafts and many others.

1.1.4.1.1 Agriculture

Over 96% of the landscape's population make their living from farming and livestock rearing, although cultivated land accounts for no more than 20% of the landscape's total area[43] . Farming is still very much in its infancy for most farmers. 95% of plantations do not use any fertiliser, even mineral or organic[44] . Fewer than 1,000 farmers have adopted agro-ecological farming methods (SCV, crop association and succession, etc.). Soil degradation has been recorded in the cultivated areas of the landscape for more than a decade, with the coastal areas (Communes of *Beheloke*, *Itampolo* and *Soalara* Sud) experiencing the worst degradation[45] . More than 80% of the region's farmland is untitled, with the remainder either sharecropped, leased or rent-free. The yield per hectare of almost all the crops grown on the plateau has fallen: manioc, maize, pulses, sweet potatoes, etc. The main causes are rainfall conditions (spatial and temporal distribution and volume), soil degradation and plant diseases, which are increasing in frequency and intensity. The technical capacity of farmers is also lacking, and although agriculture is the most widely practised economic activity, it has never developed into a professional activity.

1.1.4.1.2 Breeding

Livestock farming, i.e. the ownership of small and large livestock (goats, sheep and zebu) as well as poultry, plays a key role in the traditions and economic life of the people of the *Mahafale*

---

[43] SIRSA 2004 and SAP 2010 data

[44] Agricultural Statistics Service, MAEP, 2006

[45] FOFIFA / INSTAT / Cornell - 2001

landscape. Animals, in particular herds of zebus, determine the economic and social status of their owners.

1.1.4.1.2.1 Socio-economic importance

In the *Mahafale* area, there is no banking system. As a result, livestock represent the most important financial reserve for households. Owners of large herds of zebu[46] belong to the *"manankanàna"*, the richest social group. Breeders with a small number of zebus or a large herd of goats and sheep form a second class known as *"mahitahita"*. Further down the economic ladder, members of the third and fourth classes, the *"mahavelompo"* and *"latsa"*, have no animals at all or own only a few goats or sheep. They often work as day labourers for the first two groups.

Animals are sold if their owners need money to buy food during periods of '*kere*' (famine), to buy clothes and other necessities and, in a few cases, to finance children's education. They are also sacrificed, especially zebus, during traditional ceremonies. When the owner dies, part of the herd is sacrificed and the rest inherited by the male descendants.

The loss of a large number of animals, or even an entire herd, as a result of disease during periods of drought, theft of livestock, or simply "bad management", can lead to the downfall of a poorer social category. The smaller the herd, the greater the risk. On the other hand, there is the possibility of economic ascent through inheritance, as well as through the purchase of goats and sheep and then zebus.

1.1.4.1.2.2 Cultural importance

Goats, sheep and especially zebus are sacrificed during traditional ceremonies, including funerals and the construction of a tomb, weddings or the introduction of a new "*hazomanga*". These sacrifices have a very strong cultural and social importance, and people without livestock or money have to borrow an animal which they repay by providing an animal of the same size later.

1.1.4.1.2.3 Characterisation of breeding systems

Livestock farming systems in the *Mahafale* landscape are subject to an extremely dry climate and a lack of food and water for several months of the year. Farmers have developed strategies to adapt to these difficult conditions.

1.1.4.2 Transhumance

---

[46] In MARP REPORT, Research for Sustainable Development, SuLaMa, 2011: The importance of livestock has declined since the 1960s. Before independence (before 1960), there were more animals and almost every family had zebus. Herds could number several thousand head (up to 10,000 according to one informant). This put great pressure on the natural vegetation in the areas used for grazing and contributed to deforestation. Zebus were already the most important type of livestock and were not sold (or bartered before the introduction of money). Ten goats could be exchanged for 1 zebu to increase the size of the herd.
After the 1960s, herd sizes fell sharply as a result of droughts, which forced cattle to be sold at low prices; the reduction in grazing areas as a result of the expansion of cultivated areas; epidemics (particularly in the 1980s); and cattle theft.
For some time now, we have seen the beginnings of an intensification of farming systems, with two fodder plants being planted more frequently and over larger areas: *samata* (*Euphorbia stenoclada*), a local species that is abundant in coastal thickets and is planted in the village and around the zebu pens; and *raketa* (*Opuntia* sp.), an exotic species planted in the fences of fields and zebu pens, but also as a fodder crop in open fields.

For people living on the coast, during the dry season, i.e. between April and November, the herds graze in coastal areas and nearby villages, where they can find sufficient food and water, despite the lack of rain.

Photo 3: Transhumance to the reception area in October

At the end of this period, food became scarce and the nutritional quality of key forage plants began to decline. The quality of *samanta*[47] in particular declined, as it became indigestible due to its higher latex content. Water availability, on the other hand, does not appear to be a critical factor, despite the relatively high salinity of the water drawn from wells and natural waterholes.

To cope with the lack of fodder, almost all the zebus and even some goats and sheep move to the grazing areas to the east of the limestone plateau. This transhumance (*làlan'aomby*) takes place as soon as the first rain falls on the limestone plateau, which occurs between late October and December, a little earlier than in the coastal zone.

Traditional customs and management methods

More generally, pastoral practices cannot be separated from the overall religious and magical practices of the groups in the region. The zebu remains at the centre of communication between the living and the lineage ancestors, on whom prosperity ultimately depends. It is always by sacrificing zebus, the wealth par excellence, that the quality of this communication is ensured.

Traditionally, transhumance is governed by a set of rules which are presented below:

- *Titike* or 'social pact' between natives and transhumants.

On arrival in a host area, the transhumant must ask the local people to perform the "*titike*". During this pact, an animal (preferably a zebu) is sacrificed by the *Mpitan-kazomanga* in front of the sacred pole. The transhumant then swears to follow local rules or face divine punishment: he is forbidden to steal his neighbours' belongings, to lie, to commit adultery and to fail in his "local" obligations. In addition, the natives designate the rules for the use of the land, as well as the watering places and pastures to be used by the transhumant.

[47] Zoological Name: Euphorbia stenoclada

In addition, it was customary for the owner of a herd to visit the host village before the arrival of his herd in order to announce its arrival and define the rules of access to resources in consultation with the locals.

- The "*tsotse*" or blessing of the transhumant.

For the people of the *Mahafale* landscape, whatever their ethnicity, there is a similarity in the traditions of transhumance. Before setting off, they perform a sacrifice called *"Tsotse*[48] *"*. During the cult, the transhumants ask the blessing of the gods and ancestors for the prosperity of their livestock and for good health during the transhumance. They sacrifice a goat or sheep[49] but not an ox. On this occasion, they sprinkle the bullpen and the oxen with the "grigri of the ancestors". On the way back, they also make a sacrifice by way of thanks.

- The *Kabary* or rite of social conciliation and dispute resolution.

The *Kabary* is both a rite and a social conciliation meeting conducted by the *Mpitan-kazomanga* to settle disputes and disagreements within society. Offences committed and taboos or prohibitions flouted are discussed and settled, with the possibility of sanctions.

- The *Vilo*, or symbolic mark to assert land rights

The ox plays a key role in the conquest and occupation of the new space. It bears a symbolic mark on its ear, the *vilo*, which perpetuates and asserts the existence of a group through the ages and in society. According to this concept, the ox, by its very presence, enables the breeder to naturally assert his rights over the land.

- Other rites and taboos associated with transhumance

Other traditional rites are linked to transhumance, such as the use of grigri to ward off disease and *malaso*, and to ensure the good development and prosperity of the herd. There are also various taboos, such as the ban on wearing salt during transhumance and the use of flaming torches at night.

At "*Kialo*[50] ", a group of acquaintances immolate a goat to ask the locals for hospitality. The old transhumants take the opportunity to explain to the newcomers the rules that govern their host area.

During this regrouping, the herds are accompanied by the owners themselves, or by members of their family or villagers hired to supply them with water and fodder. These people will stay with the animals throughout the transhumance period, i.e. for up to five or six consecutive months.

In the eastern savannahs, the herds rest and fatten up, feeding on abundant, good-quality fodder plants. The quantity of water, although a limiting factor at other times of the year, is generally sufficient to support not only the local herds but also those on the coast. At the end of the rainy season, water becomes increasingly scarce and the herds return to the coast in March or April, thus completing the annual cycle.

[48] Tsotse = Tso-drano or blessing

[49] The sacrifice of a chicken is acceptable if you don't have goats or sheep for those who don't eat goats or sheep.

[50] The reception area for transhumants

The *konda*[51] , or cart-drawn zebus, do not make the transhumance and remain all year round in the coastal zone. They are partly fed on crop residues, which provide a good but not abundant fodder supplement. Sometimes they are staked inside the enclosures *(vala)* after the harvest.

Most goats and sheep remain in the coastal area throughout t h e year. Only a few, mainly owned by zebu farmers who are unable to hire a salaried employee to look after their small ruminants, take part in transhumance.

The most widespread diseases in the *Mahafale* landscape are the following:

- *Besoroke* or "coal", which causes inflammation of the shoulder;
- *Beareke*, a fatal disease that swells the pancreas in animals;
- *Kopake*, which causes paralysis.

The *Mahafale* like to breed red or white oxen because they are easy to sell in this way. Moreover, today's breeders are beginning to focus on the transaction (sale of oxen, bartering of draught oxen). They are more concerned about the health of their livestock and agree to have their oxen vaccinated, even if vaccination costs 1000 *Ariary* per zebu.

Dark-coloured oxen, such as black or brown, are used to worship Tambahoake (a spirit considered to own the forest). To respect Tambahoake, the possessed should immolate an ox, but not a goat or sheep.

To make a sacrifice for rain to fall, a black ox with a white forehead (*Tsiriry in Mahafale*) is offered. In short, dark-coloured oxen are used for offerings.

According to our findings, respondents in the *Betioky* district have many more cattle than those in the *Ampanihy* district. They fear the possible disappearance of the Malagasy cattle breed.

The transhumants revealed that only the heteropogon contortus[52] (*ahidambo in mg*) is preferred by the oxen, as it fattens them quickly and, above all, has aphrodisiac powers. Samanta[53] and *raketa*[54] (burnt thorn cactus) are survival foods.

Because of the insecurity, the transhumant herders are taking refuge inside the park. They spread out over the southern part of the new extension.

According to our brief interview with the Park's sector chiefs, the sector chief based in *Beahitse* informed us of the existence of the "Ten Commandments of *Itampolo*", which state that the roaming of oxen within the Park, the existence of hamlets within the Park, burial within the Park and the cutting of coffin wood within the Park are all permitted. Within the park, *Ankororoke* is a well-known resting place for transhumants coming from the southern part of *Itampolo*.

The establishment of rest areas outside the park merits careful consideration, given that transhumant herders currently travel inside the park because of the insecurity.

---

[51] This word comes from the French condamner, because the draught oxen are condemned to milk the carts. This is their only role.

[52] Some members of the Zamasy CoBa have already tried to grow this variety of grass at home in preparation for stopping transhumance at a later date, because for them the search for this grass means having to travel a long way.

[53] samanta (Euphorbia stenoclada)

[54] Raketa (Opuntia sp.).

For information, they have already set up a draft *Dina* to govern the *Antselempasy* area, in the *Maroarivo* district of *Betioky*.

1.1.4.2.1 Workforce

In general, the herds are looked after by young men from the owner's family. When a family has no male children but has the money, it may hire a professional herdsman to do the job. In this case, the employee receives a 2-year-old zebu as an annual salary.

At the start of the dry season, the workload involved in herding zebu cattle is less. Cows need to be milked once a day, and herds are easy to keep an eye on. As the season progresses, the need for labour increases due to the use of samanta and *raketa* as fodder plants. This use peaks in August and September. During this period, the herdsmen have to work almost all day to harvest and prepare this fodder. The samanta branches *are* cut and chopped, while the *raketa* leaves are burnt superficially to remove the thorns before being chopped into small pieces. Collecting the firewood, gathering the *raketa* stems, arranging them in piles, burning the thorns (done at night) and then chopping them up take up the entire nights and days of a person looking after 30 zebus.

At the end of the dry season, the cows no longer need to be milked in order to save the milk for the calves. The workload involved in preparing fodder plants increases. Watching over the herds then becomes the most demanding job, as transhumance requires extra labour to accompany the herds. Guarding the herds to the east of the plateau also represents a heavy burden because of the risk of cattle theft, even if this risk is reduced during the rainy season.

The herds of goats and sheep are looked after by young boys from the age of 7 or 8. The goats are milked 3 times a day during the rainy season until the start of the dry season.

Finally, the livestock sector (particularly the zebu sector) has evolved very little, having remained at the "contemplative" stage, more cultural than economic. It has been badly affected by the unfavourable climatic conditions of recent years, as the long droughts are synonymous with a lack of fodder, watering problems, long transhumance (distance and duration) and more frequent and intense animal diseases. In addition, insecurity, correlated with drought, famine and political instability, continues to worsen. Fodder crops are grown by very few farmers, and have not yet reached a viable scale.

1.1.4.3 Craftsmanship

Craft activities include basketry and weaving (*tegnone*), sculpture (*Mahafale* art, *aloalo*), *Mahafale* tomb building, cart making, forging, carpentry, house building and jewellery. Generally speaking, crafts in the *Mahafale* landscape are characterised by a lack of organisation and professionalism, under-equipment, insufficient and expensive raw materials, a lack of outlets and low production.

60% of *Mahafale* women know how to weave mats and baskets from fibres for their family needs, but this activity has become a sideline, driven by the need to make up insufficient income for family expenses: purchase of clothing and medicines, social security contributions, purchase of basic necessities, etc. A handicap of this profession is the lack of raw materials.

Cart-making and the forging of small agricultural implements (spades, axes, assegais) play an important role in income-generating activities. The main outlet is the weekly market.

The third most common activity is house-building and, above all, changing the roof to *"boka*[55] *"* during the dry season. This is a regular activity and provides a relatively important source of income for some young people during the lean season. Other craft activities are very limited.

The *Mahafale* landscape is known the world over for its mohair carpet weaving. Carving is practised by everyone, with the exception of *aloalo* carving[56] , which is reserved exclusively for men, and cart building, which is the means of transport most used by the people of the landscape. *Aloalo* carving and cart building are beneficial to those involved in this sector.

On the other hand, during the harvest period, there is an interruption in production among craftsmen.

But at present, carpentry, embroidery and silk weaving are beginning to gain ground.

Finally, small non-agricultural businesses, such as the manufacture of local rum (*toaka gasy*), can also be considered as small-scale production.

1.2 Cohesion within the community

The starting point for an assessment of citizen participation in forest resource conservation actions is the relevance of the scale of local development action, i.e. the commune itself. The NGO leaders stated *that "it is easier to create solidarity in the countryside than in the city".* However, each commune or territory of the elements of the population of the commune and the overall reality in the rural community will thus be justified, to see if this has any consequence whatsoever on the participation of each person in the development of their community or the conservation of biodiversity. Can we then achieve the participation required by democracy if there is social differentiation?

To this end, a few points will be raised to clarify the situation.

1.2.1 Family cohesion

The family is the basic unit of society. It represents an ideal organisation for effective, well-coordinated participation. In fact, the incentive to mobilise the family is made easier by the rules and disciplines that bind family members closely together.

Furthermore, the Malagasy family is still a traditional institution where all collective representations and customs are geared towards reproducing the *"aina"*, warming up to the idea of living together and passing on a material, biological and symbolic heritage from generation to generation. In this way, the family group can exercise the strictest control over its members, because within the group, the behaviour of family members is programmed according to this institution.

The reality observed in the field reflects family mutual aid incited by the heads of the family (male or female) or mutual aid due to affinities developed in the past. In general, this mutual aid is called "*rima*" in *Mahafale* or "*valin-tanana*". Family relationships therefore make it easier to

---

[55] Dry Heteropogon contortus

[56] Sculpture intended to decorate a tomb in the region.

mobilise the population, as affinities enable the information needed to make everyone aware of their duty and encourage them to do it to be passed on more quickly.

However, deep down in this reality of the *Mahafale* plateau, the family institution is tending to lose its virtue. When it comes to conserving renewable natural resources, the members are no longer concerned about observing the rules for the sake of the group and their own interests, because new values are now emerging and they have more freedom of choice. As a result, the family itself is beginning to break up and the charisma of those in authority is under threat. The implementation of the management transfer around the *Mahafale* plateau is tearing apart the social fabric. The case of CoBa *Mizakamasy* and *Ambolisogno* proves this point.

These family groupings distinguish themselves from one another and rekindle divisions between them: the groups of the first arrivals want to distinguish themselves from the new arrivals. Moreover, identification with one's ancestors leads to divisions within the community. The descendants of the large families who founded the terroirs are proud to be the *"Tompon-tany"* (owner of the terroir) and as a result, there are latent disagreements between the first arrivals and the new arrivals, known as neo-rurals.

The "neo-rurals" are emerging because they are better educated than the first arrivals, i.e. the natives; the neo-rurals administer the plateau in their own way. They are teachers and farmers living in the communes of the *Mahafale* plateau.

During an investigation carried out during our field visit, people from the large families who had founded the territories where forest resource management was transferred to the communes told us that when their management tools were being drawn up, those in power, together with the environmental NGOs, chose who could and who could not participate as they saw fit. They unanimously declared that there was nepotism in the constitution of the management committees that were to participate in the development o f these management tools and in the management of natural resources.

On the other hand, during our interview with the district chiefs in the commune, they stated that the development of these management tools requires a minimum of capacity:

> *"You can't design an effective and appropriate development strategy or land-use plan if you're not educated. [Bemanateza* district chief]

So they have challenged people's intellectual level. However, the partners are catalysts and their role is to help people formulate their aspirations in the development plan and conservation plan for these natural resources.

In the *Mahafale* landscape, the neo-ruralists want to change the political situation and above all the existing social structure. They try to marginalise the natives in a variety of ways, particularly in public affairs and official training meetings organised by support NGOs. As a result, there is a breakdown in cohesion in the commune, caused by hidden social antagonism: hence the low rate of participation in the implementation of the development plan and forest resource management tools.

Nonetheless, the communities of the *Mahafale* landscape have generally expressed their desire to conserve the forest resources around them.

Figure 1: Motivation to protect the forest

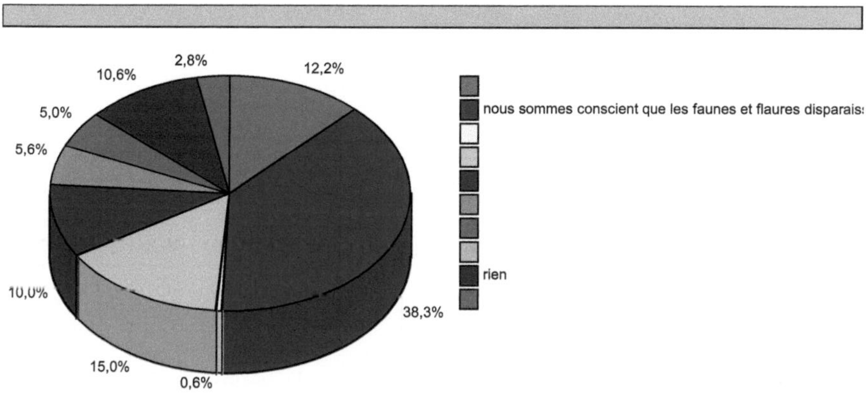

Source: Our own survey, 2014

In this graph, three reasons motivate communities to protect the forest resources around them. The first reason is the community's awareness of the threat of disappearance of flora and fauna: 38.3%. The second reason for motivating people to protect the forest is due to the awareness-raising activities of the WWF: 15% of respondents. The last reason is fear of the aridity of their soil: 12.2% of survey participants. On the other hand, 10.6% of respondents had no motivation to protect the forest resources around *Tsimanampesotse* Park. 10% were motivated to protect their forests in order to obtain financial aid from support organisations.

1.2.2 Belonging to a social rank

Belonging to a traditional social order is more or less hidden. It has little influence on people's daily lives. We have already mentioned this in previous chapters.

A person's rank is defined above all by their feelings of identity, their reticence or their affinities.

The table below shows some of the key features of people's participation by rank.

Table 3: Level of participation by rank

Level of participation	**Local Authorities**				**VOI members**				**Non-members of the VOI**				**Grand total**	
Type	Men		Woman		Men		Woman		Men		Woman			
Social rank	Nb	%	Nb	%	Nb	%	Nb	%	Nb	%	Nb	%	Nb	%
Andriana	84	47,19	-	-	17	10,11	08	25	45	16,98	62	39,49	216	27
Hova	58	32,58	-	-	78	46,42	15	46,87	158	60,37	75	47,77	384	48
Andevo	0	0	0	0	0	0	0	0	0	0	0	0	0	0
Foreigners	36	20,22	-	-	76	45,23	9	28,12	59	22,64	20	12,73	200	25
Total	178	100	-	-	171	100	32	100	262	100	157	100	800	100

Source: Our own survey, 2014

According to this table, it is the *Hova* who are most motivated to participate in the management of renewable natural resources, as they number 384 people and their participation rate in the transfer of natural resource management is 48%. This is followed by the *Andriana* and foreigners (27% versus 25%).

In addition, *Andriana* descendants are much more interested in the development of their territory than Hova descendants (47.19% compared to 32.58%). This may be due to the high number of *Andriana* on the executive board of each commune.

It should also be noted that the *Mahafale* community was divided into only two orders: *Andriana* and *hova*. The *andevo* group did not exist among the Mahafale.

With regard to the degree of participation by gender, that of women is very low compared with that of men, since 23.62% of the total population are women and only 4% of them have participated in the management of forest resources. This attitude can be explained by their capacity as well as their culture, which limits them to work in the home. Apart from the *Andriana* who dare to assert themselves, it is difficult to ask respondents to define themselves in terms of their social rank, and it is through their behaviour during the interview and their profits that we have drawn conclusions.

The *Andriana* administer their territory with the neo-rurals. As for the *Hova*, they feel left out of this administration because the people with the greatest responsibility in their territory are *Andriana* and this is why they dare to claim the status of landowners. The *Hova* descendants are totally marginalised from the executive office and even in the *Fokontany*, but they do the work in the fields as salaried employees.

This form of exclusion, based on a virtual social division, is blocking the participatory approach to the successful development of the *Mahafale* landscape.

During a group interview with the people of the *Mahafale* plateau, we asked them about their perceptions of cohabitation and relations with other residents of different social standing, from which we deduced that this is proving difficult because there is a risk of total destruction of social cohesion.

The "class struggle" between the non-natives and the natives of the *Mahafale* landscape is leading to the destructuring of the community, which in turn will demotivate the citizens in the development process and especially in the process of managing the forest ecosystem.

Nevertheless, the marriage between *Hova* and *Andriana* led to a mixing of the Communes. This intermingling had two consequences: it reduced the tenacity of social antagonism, but it also strengthened the cohesion of the community so that it could "join hands" (*mifanome-tanana*) for the conservation of natural resources and the development of their commune. The non-participation reported among the *Hova* descendants can be transformed into indirect participation, as they are the ones who provide the workforce in their commune.

In short, we can expect the result of this intermingling to appear soon, but until now social rank has been decisive in the division of tasks. This could disrupt women's cohesion, given that everyone's role is already established from one family to another.

1.2.3 The existence of strong institutions in the municipalities

1.2.3.1 The temples

"Toy ny ladim-boatavo ny olombelona ka raha fotorina dia iray ihany" - Literally, it means that man is considered to be the creeping stem of a gourd, but when you look for its root, it's unique. This adage illustrates the Malagasy *fihavanana* based on mutual aid and living together. People are aware of their source, and that source is God or the ancestors:

> *"Moreover, the razana are not passive; they are supposed to intervene in the daily lives of the living, either penalising those who fail to show the respect and veneration to which they are entitled, or, on the contrary, rewarding those who fulfil their obligations of all kinds towards them"* [RAKOTO (I) ; RAMIANDRASOA (F) et RANDRIAMBOAVONJY (R), 1995].

There is always a link between the living and the ancestors.

When local people identify with their noble ancestors, the founders of their land, it seems that they are attached to the temple. The fact of being a descendant of a great family, of attending the temple, of having an important responsibility there, pushes the individual to be dynamic and active in serving his territory.

The inhabitants of the *Mahafale* plateau are very attached to their traditional religion. This attachment is also reflected in the first contact the leaders of the *Marofototse* and *Behombe* districts had with the NGO AVSF when it launched the introduction of a new cultivation technique: SCV in 2007.

The plateau is home to many sacred places. These are generally the homes of *Tambahoake*, a spirit worshipped by the communities. In fact, most of the sacred places are included in the land transferred to the communities.

Generally speaking, sects do not act as catalysts for development, but rather as preachers of the Bible. However, they are changing their approach. They let their followers participate in the conservation of forest resources in order to attract many supporters.

During our interview with a *Jesosy Famonjena* catechist, he stated that

> *"... even if there is nothing to hope for in this earthly world, we encourage people to respect and protect natural resources so that they are aware that only God can give them new hope: a life without hardship or tears that is eternal life".*

This proves the despair of the Malagasy people as far as economic growth is concerned, and their last hope is to entrust their lives to the Creator; they are no longer looking for a way out of their poverty; moreover, the intellectual level of these followers is generally very low. Moreover, these sects make every effort to gain a large number of followers. To do this, they recruit followers from the main churches, both Catholic and Protestant, hence the latent disagreement between the leaders of the temples in the commune of *Betioky*. Instead of being a unifying force that encourages community participation, the temple has become a source of community division. The sects are becoming marginalised and taking root on the outskirts of the commune.

Finally, the notion of religious community is omnipresent on the *Mahafale* plateau. It is of great importance because it structures the social rhythm, and because it brings together almost the entire population according to various symbolic temporalities. Whether Catholic, Protestant or traditional, the *mpitan-kazomanga* (master of worship) has great influence, and the relationship with God is both submissive (God is the higher power who orders the world) and dynamic (people implore God or his representatives, interacting with the invisible world). However, the bonds of solidarity between believers are only weakly developed, each having a personal relationship with God. So religious community is not synonymous with solidarity.

Only certain traditional rites, such as the *enga lolo* (funeral ceremony), involve symbolic exchanges and economic reciprocity at family level. For example, during certain ceremonies, gifts are distributed to the neediest members of the kinship group, through donations to the spirits (*soro*). In this way, the satisfied spirits fulfil the wishes of the donors, and the people who receive the gifts (in kind) undertake to render services to their benefactors.

1.2.3.2 The strong presence of a political party

Political parties and society are like water and rice, they are interdependent. Parties play a fundamental role in the formation of political will in the country. They provide the State or decentralised local authorities with political staff, through democratic elections and appointments. They must encourage the active participation of citizens in political life, in order to achieve sustainable and harmonious development in this country.

However, they are seen as a source of political unrest and conflict, corruption and financial scandal, particularly after the quest for money by CoBa president *Magnasoa Tane* from *Behombe*. This person wears two hats because he is also a member of the *Hiaraka Isika* political bureau. To take advantage of the schooners for fencing off fields or for ox pens, you have to pay 200,000 Ariary. The farmers paid this amount, but got nothing.

What's more, the CoBa contract renewal period coincided with the aborted municipal election period during the transition in 2011-2012.

Most of the CoBa management committees were members of the AVI party, and during the presidential propaganda campaign, the AVI was split into two blocs; one remained loyal to its party, which supported the candidate *Hery Rajaonarimampianina*, and the other strengthened the ranks of the new party: the MMM. Those in power chose who should participate in the transfer of forest resource management to their territories.

The crisis of 2009, caused by the protest of the TGV and its allies who wanted at all costs to validate the respect of the citizens' choice, worsened the zizanie within the supporters of the two clans, *pro-Ravalomanana* and pro-TGV.

Generally speaking, the origin of any local development initiative is dissatisfaction with a situation that no longer meets the needs and aspirations of individuals or groups of individuals within a community. This phase enables the various stakeholders to become aware of the usefulness of their participation in the development of their locality. Implementing a local development policy requires the stakeholders concerned to work together, and to set up networks and partnership mechanisms. Consequently, there can be no local development without awareness on the part of local elected representatives, a shared desire to take action, a collective capacity to launch and support the development process, the enhancement of resources, and the recognition and support of local initiatives by local representatives. In other words, raising the awareness of the population concerned to stimulate local initiatives and provoke a self-analysis of the situation by the local population with a view to reactivating an internal dynamic of discussion and change.

As a result, the event that triggers awareness is either internal or external to the community.

In the first case, the community is not the initiator of the collective awareness: the triggering event is provoked by an external actor. The initiators express their opinions and try to rally a section of the population around them in order to reinforce their position. It is a response to a situation where the solution lies with an authority outside the group concerned. In this case, when the community is in a position to react, it adopts a passive mode of action.

In the second case, the community or a small group within the community has become aware of certain problems that are undermining the population, or of certain warning signs of crises. In this second scenario, the triggering event will take place in an active mode. The community will be in control of the choice and conduct of appropriate responses to the situation. The two main mechanisms of this mode are: participation, i.e. the access of individuals to the decision-making spheres, and planning, i.e. the definition of precise objectives and the means to achieve them.

In this second case, the only prospect is that of coordinated action, carried out by people who feel concerned, motivated people who are capable of launching and supporting a development project or a project for the conservation of renewable natural resources. This is the ideal route to local development.

However, the population's participation in political life will encounter many problems in the life of the community. It directly undermines community cohesion. The population became distrustful and this affected their solidarity. *Mahafale* community life gradually degenerated. This situation became very apparent during the social mobilisation.

In short, the failure to integrate the whole population for social mobilisation due to the existence of family ties from different categories or segregation of groups in the communes, has a negative impact on people's willingness to participate in the management/conservation of forest resources and especially in community development. Moreover, only the large institutions maintain a certain level of dynamism among the residents of the communes and are able to mobilise residents more easily.

Thus, in the case of the *Mahafale* plateau, political, social and family factors have a direct influence on the participation of community members in the conservation of renewable natural resources and the development of the communes that make up the landscape.

1.2.3.3 Contradiction between two villages: ascending and descending

First of all, the *Mahafale hazomanga* community is a fundamental state within the state. The *Mahafale* do not take their *hazomanga* with them when they immigrate. They systematically return to circumcise male children or celebrate family rites in front of the *hazomanga*. The hazomanga, which is planted in the ground, is a sacred foundation. It marks ownership of the land. The founding fathers planted it to signify that they were settling there and that the land belonged to them; they made it sacred by this foundation rite, hence the very special name of camp or *Toby* (the group's land). For the inhabitants of the *Mahafale* plateau, the *hazomanga* pole and the community are inseparable, because the cosmic pole planted signifies a concrete temple of the community. The community is a praying community founded in the ground after the founding rite by the founding fathers. The pole represents the reality at the centre of the community (the life of descent). The life of descent is the heart of this system. To touch the *hazomanga* is to touch this life and therefore to kill the community. The pointed shape of the pole raised towards the sky is a continuous prayer.

The *Mahafale* community practises self-management[57] from the top down: self-management is defined as the practice of administering and managing a community in a way that allows it to take control of itself, as opposed to a "domesticating" or "despoiling" administration.

However, the town also has something to say about its management. The following figure shows the contradiction between the villages.

---

[57] Self-management of offspring can be summed up in three aspects:

- the descendants' awareness of being a community born of the same divine "aina" of descent. This community is then faced with the battle of life;
- Managing descendants also means managing the self-management of this life of descendants. The community is aware of being self-dynamic and self-responsible at both personal and community level. This is the core of downline self-management, a bottom-up self-management. This bottom-up dynamic of self-management has now been broken by missionaries and NGOs;
- Finally, self-management of descendants concerns the management of the economy of descendants (management of the subjects of descendants or 'aina' subjects of life in order to make a valid posterity or 'vagnogne' and parental solidarity of mutual aid or 'rima' at the level of lineage work).

Figure 2: Contradiction between descending and ascending villages

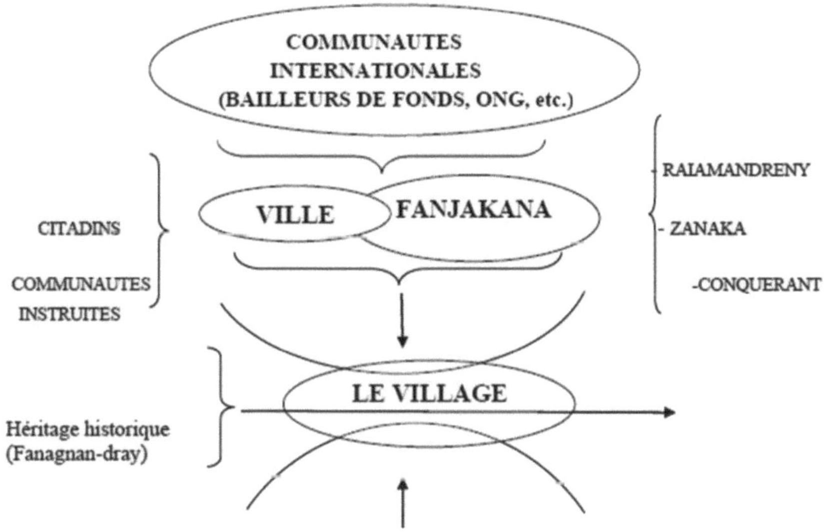

Priestly and Protective Community of Descendants (CPPD)

Source: Parson RAMBINIZANDRY, 2014

This figure shows two dynamics that clash in today's *Mahafale* village:

- bottom-up dynamics based on *hazomanga* dynamics. In this respect, the ascendant village constitutes the first level of a new historical way of living together. This will be the ascendant *fokonolona*;
- on the other hand, the *Fanjakana*, the city and the international communities that are inseparable from it penetrate and conquer this ascendant village and place their agents there. As a result of this conquest, there is a fundamental contradiction between the village and *Fanjakana.* In concrete terms, the descending village is found in the village in the form of state agents: the *Fokontany* chief, the mayor, the district chief, the gendarmerie, the extension workers and Cantonnement chiefs, teachers and the radio. On the other hand, the town is present through the merchants, but also the retired people, the educated children of the village and above all the cultural models from the town. Finally, the international communities, via the World Bank and international development NGOs, civilise the villages by spreading the messages of integrated conservation, equitable benefit sharing and sustainable management of common property resources. As a result, it is the outside agents who decide the fate of the village. On the other hand, the villagers, who are dependent on them, are marginalised, sidelined and structurally excluded. From these elements, the *Mahafale* village presents itself in its very precarious, very harsh but very essential daily life.

Internal and external domination produce a generalised crisis in the village. This crisis manifests itself in various forms of contestation. To counter this downward pressure, the communities of the *Mahafale* plateau have adopted the following two strategies:

- a strategy of refusal and withdrawal: the village defends itself in its own way, in silence and by withdrawing into itself. The community rejects commercial exchange, and is self-sufficient in terms of its descendants and self-sufficiency, hence the rejection of Christianity, the central or decentralised *Fanjakana*, the gendarme, the canton chief and, above all, the school;
- a strategy of resorting to phenomena of possession. This is the case with the *'bilo'* and *'tromba'* phenomena, which mobilise the village almost every night (a nocturnal liturgy);
- a departure strategy: in fact, the community is condemned to find its subsistence and money. The village is impoverished, the descendants have immense needs with regard to ancestors or funerals, the *Fanjakana* relationship, church, hence the departure of community members:
    - simple salaried departure or seasonal migration in search of money ;
    - longer periods of schooling and accumulation in towns and regions.

1.3 Trust between managers and employees

"If the leader persists in his negative attitudes and his refusal to trust anyone, there is no hope for him of making the law prevail; nothing is built without trust (...trust is built)" [DEMONQUE M., EICHENDERGER J.Y, 1968].

This requires mutual trust between managers and those being managed; it's a process that is built up little by little.

The participation of each member of the group depends on the choice of leader, and his or her ability to ensure that mutual trust or the commune's power relationship prevails. Leaders who want to involve the whole community in a project must show that they are free of any commitment, that they are not stifled by external obligations or overly concerned only with their own interests, because there are other concerns too: the general interest of the area or the commune. It is in this way that they will be able to gain the trust of the citizens, and this can be confirmed because the socio-political relationship of the *Mahafale* plateau influences the forms of participation of the inhabitants.

Relations between the authorities and the local population vary from one social group to another. The non-existence of a structure for dialogue at Commune level in which citizens can express themselves freely may be the main reason for this. Despite the Commune's support for the creation of grassroots communities to manage natural resources and the revitalisation of farmers' organisations, the Commune's objective is still to seek collaboration and funding. It does not yet intend to set up a platform that could be a forum for dialogue between the authorities and the local population, because of the lack of resources but also, perhaps, because it is not its priority.

The development committee that was supposed to be set up to represent the various categories of rural stakeholders is not functional because of these non-representative structures, and the authorities have no desire to reform and revitalise it. In addition, there is a lack of willingness on the part of the population to participate, in the belief that communal affairs belong solely to the leaders, and that it is up to them to ensure the well-being of every citizen.

During a debate, for example, a farmer said: *"It was the President himself who said that tomorrow everyone will have at least a 4L car and will be rich, but until now my situation has only got worse"*.

It's a statement, gratuitous and vulgar perhaps, but it reflects a certain false image of those in power and also reflects the attitudes of citizens towards the authorities.

To confirm this spirit of expectation, during the consultation of the various grassroots communities, many of them thought that having regular situations such as statutes and internal regulations automatically led to funding.

Furthermore, instability, especially during the transition period, is felt in the countryside to be the result of the State losing its way, leading to a loss of the value of Malagasy regulations and leading the State into anarchy, as *"the current State has no regulations"*, and the transhumants of the *Mahafale* plateau, showing their lack of confidence in the leaders, consider that the increase in the *malaso* phenomenon is a manoeuvre by the State leaders. "*Cattle rearing on the Mahafale plateau has deteriorated, it's the state that's causing the malaso to reign*", remarked one transhumant during an interview. They accused the leaders of creating this *malaso* phenomenon and mocked the central authorities for their inability to resolve the problem of insecurity, which is undermining the nation today. As a result, some transhumants are afraid that the Malagasy cattle breed will dissolve because of the phenomenon of massive exports and theft of oxen. *"The State must keep a close eye on cattle farming, otherwise there will be no more cattle"*.

They accuse those in power (central or decentralised) of selling their land. For them, the transfer of natural resource management is just a hidden form of selling off the lands of their *'tanin-draza'* ancestors.

*"...since the rise of the Ambaniandro to power, we've heard that there's been a proliferation of land grabs and sales here and there. Why are they selling land that doesn't belong to them?... They've even demarcated our "kiririsa" (a place close to the house where the children can play together) to make it part of the Tsimanampesotse, which is 70km from where we live. We've been asking for this for a long time, but I think our children will do it. [*Interview with a prominent person in *Behombe*].

According to this prominent figure, they no longer have the strength to lay direct claim to the land of their ancestors, perhaps because of the modern political system. He believes that their descendants will no longer easily accept this situation. However, in an indirect way, the notables of the *Mahafale* plateau are countering this policy or "domination" from outside by allowing members of the community to clear the *Tsimanampesotse* park. In addition, there was a communication problem during the public consultation. There was little explanation of the reasons for extending the park, but they were quick to collect community signatures to authenticate their cases.

The following diagram summarises the communication problems between managers and employees.

Figure 3: Communication problems between elected representatives and the local population

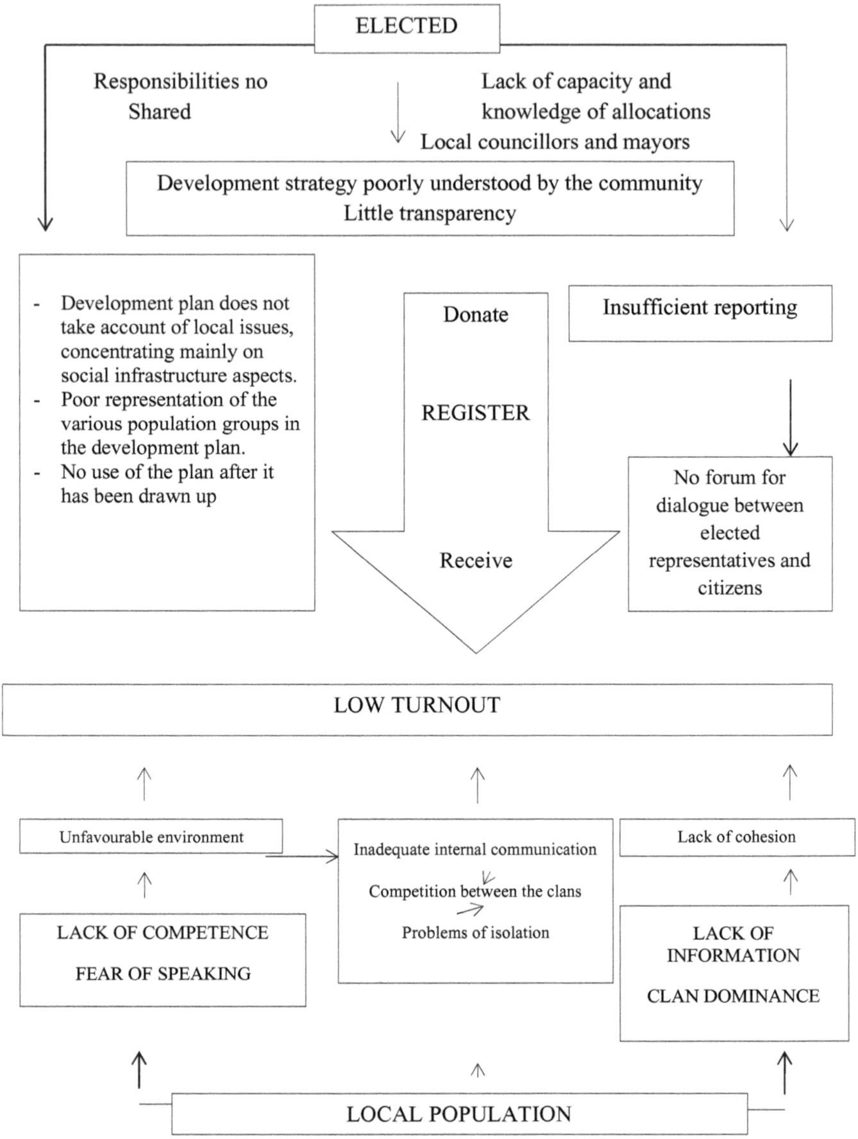

Source: RAMBINIZANDRY, 2014

This diagram reveals the problem of communication between managers and employees. In this diagram, the low level of participation by those being led is due to a lack of competence and cohesion as a result of competition between the clans, as well as inadequate internal communication and insufficient information circulating within the population. This lack of communication means that the population is afraid to speak out. On the other hand, elected leaders are very reluctant to share their responsibilities. Because of a lack of capacity and a lack of knowledge of their remit, leaders do not pay attention to citizens. As a result, they are not motivated to set up a forum for dialogue between themselves and citizens. As a result, there is little or no reporting to the community.

1.3.1 Influence of Community values and institutions

We cannot hastily conclude that these alone can guide the way in which citizens participate in the *Mahafale* plateau. Even if it has been shown that the temple creates solidarity and political party politics engender reticence and the non-integration of the community, it is also necessary to focus on the relationships between the leaders and leaders of the communes in order to create strong *Fokontany* institutions that will condition mobilisation.

1.3.1.1 *Fokontany* bureaucracy

If the neighbourhood leader really inspires confidence in the members and does not simply represent a formal authority according to the hierarchical structure, but asserts himself as a leader with natural charisma, he will have more influence over these members.

The image played by the district manager and his outlook can be associated with that of the manager in today's democratic style of leadership. He has to plan his work according to what his colleagues are likely to accept. He uses tact and diplomacy in assigning responsibilities and ensuring that his staff accept what he asks of them. He checks and monitors the progress made by his staff so as to be able to make a good projection.

The members of the district office form an administrative structure conducive to communication, as they have communication and information officers from each hamlet for the *Fokontany.* These agents, who are closer to the population, have a representative on the district office. However, the members of the office are not really up to their tasks. They don't know how to get residents involved. Their leadership skills are not enough to help the group achieve its objectives. If these residents are not making progress, it is because the leader has not been able to play his role in motivating his team.

In addition, the hamlets that make up a district are quite far apart, which disrupts communication between hamlets and districts. The political context also plays a major role. The population's relationship with the neighbourhood chief is indeed a very sensitive issue, as the ambivalence of his status is seen as one of the obstacles to participation.

The district chief is appointed at the whim of the mayor, but is not elected through a democratic process. Citizens have less and less confidence in the state's representatives in the district, so they take the time to get to know them better before placing their trust in the appointed leaders, or even those parachuted in from the commune's capital.

In fact, some district chiefs from the outskirts of the communes were parachuted in, as they had only lived there for a month before the communal election, with the aim of being appointed district chiefs of these localities. They are therefore foreigners who took advantage of the election to integrate into the neighbourhoods.

In order to better manage and control the population, they are obliged to resort to different kinds of *"dina"* (rules or conventions for disputes). They created these conventions themselves and validated them with a few village citizens. The application of these *dina* was therefore met with reluctance by the members of the neighbourhood office, because the general assembly that decided to set them up was not really representative of all the neighbourhood's inhabitants, but their effectiveness was appreciated. This raises the question of whether or not neighbourhood leaders should be excluded from the conservation and/or natural resource management project and the development project in order to better encourage and maximise citizen participation.

The majority of residents feel marginalised when it comes to decision-making; moreover, they feel very little involvement in the major decisions to be taken in the neighbourhoods, including decisions on which *dina* to apply and which terroirs to transfer to a CoBa. The stakeholders spoke about this during the interviews, and we felt that they were taking advantage of this habitual apathy on the part of the passive population to reinforce a personal strategy to seize power. They even told them that *"the residents are just waiters and they are afraid to take responsibility".*

What's more, it's difficult for local people to prepare administrative paperwork at the offices, because there are no official premises; in fact, the home of the neighbourhood chief is generally used as the neighbourhood office. Fear of state institutions and politics still persists. In addition, there are different political tendencies; the political opponents of the district chief have difficulty obtaining any administrative paperwork.

The villagers are not motivated to take part in community actions organised by the leaders, particularly development projects, because not everyone feels that they are the direct beneficiaries of the projects and works, i.e. they think that these works can be used for electoral purposes. Faced with such a situation, it is difficult to turn this population into a dynamic driving force for the neighbourhoods. You have to gain their trust and appeal to their initiative, otherwise political parties will take over and the targeted members of the community will become increasingly marginalised.

In conclusion, the communes and neighbourhood offices of the *Mahafale* plateau are an important entity for the continuity of community action in the neighbourhoods; however, based on this observation, we doubt that the relations between those in power in the neighbourhoods and the citizens can work. Contact with those in power in the state still frightens the population, as does anything to do with politics. A large part of the population is afraid of institutions, especially when it comes to administrative procedures. This attitude can be illustrated by the rate of possession of birth certificates in the communes, which is still less than half the population. Yet it is dialogue between the population and the institutions that strengthens the motivation of citizens to participate in all community activities.

1.3.1.2 Highlighting the gender approach

Thanks to democracy and the rule of law, women's rights have long been established in theory and progress has been made; nevertheless, sustainable human development still requires major efforts.

Indeed, although women are now beginning to be considered for participation in decision-making, they are not yet able to play important roles. There is still a need to create a more favourable environment for women's participation in political and economic affairs, particularly at grassroots, local community level.

The division of labour between men and women has changed considerably in Madagascar's major cities. Men and women can respect each other, cooperate and help each other to improve their immediate environment. In general, women's potential for development has not been blocked or taken lightly. However, women who do farm work and household chores are not well regarded, as this is not accepted as a profession. Women also contribute to the household budget, and their timetable is as full as that of men, but they are still more or less undervalued, despite the productive and social value of their actions for local development.

In the *Mahafale* landscape, women are generally excluded from public affairs, which remain the prerogative of men, especially middle-aged men. For example, there are no women on the local development committee. The population has not understood the value of women's participation in decision-making on the future of a commune; yet, given the tasks to which they are assigned, it would seem that they represent a lever for community development.

Table 4: Level of participation by gender

Degree of participation	Local Authorities				VOI members				Non VOI members				Grand total	
Type	Men		Woman		Men		Woman		Men		Woman			
Origin	Nb	%	Nb	%	Nb	%	Nb	%	Nb	%	Nb	%	Nb	%
Native	142	79,77	-	-	95	55,55	23	71,87	203	77,48	137	87,26	600	75
Foreigners	36	20,22	-	-	76	44,44	9	28,12	59	21,51	20	12,73	200	25
Total	178	100	-	-	171	100	32	100	262	100	157	100	800	100

Source: Our own survey, 2014

This table shows the degree of participation of the citizens of the *Mahafale* plateau according to gender in the process of developing a forest resource management tool, in the conservation of these forest resources and in the coordination of development actions. Thus 87.26% of the female population surveyed could not participate in the management and conservation of forest resources, compared with 77.48% of the male population. Women at least participate in the process of developing the planning and management tool for local renewable natural resources. This non-participation can be explained by the higher illiteracy rate of women compared with men; moreover, women are preoccupied with their domestic schedule, so they have no time to devote to community activities and environmental protection; finally, the division of tasks between men and women is still in force in the *Mahafale* landscape and, as a result, women do not yet have any decision-making power. This is demonstrated by the difference between the participation rate of women according to their origin; in fact, 20 out of 29 women not from the region, i.e. 68.96%, did not participate; on the other hand, 137 out of 160 women from the region, i.e. 85.62%, could not participate despite the use of the participatory approach to develop this tool. In fact, a participatory approach is based on partnership between all development stakeholders, but it obviously takes into account the plurality of decision-making levels.

1.3.1.3 The power of the Raiamandreny

Malagasy culture has marked society as a community governed by elders, and knowledge of this fact is fundamental to the study of power in Madagascar. This was previously noticeable in the constitution of the family, where the hierarchy in the sharing of power depended first and foremost on the man, because Malagasy families are patriarchal. It is the character of the power of the elders that we wish to evoke in this section, but not a gerontocracy in the sense commonly used by politicians.

This form of traditional power, based on the power of men and especially parents, seems to be perpetuated from generation to generation. If this situation continues, there is no doubt that young people and women are excluded from responsibility for the life of the region, as the myth of experienced elders who are able to take decisions persists in family relationships and among the vital forces who will participate in community life. Young people, who are considered to be the potential and future of the *Mahafale* plateau, are not integrated enough to really be part of the working population or to be leaders in their community. They are often excluded from major development projects and from the development of a forest resource management tool for their area. Their lack of education, due to the high cost of education, partly explains their passivity, or rather their lack of initiative. Their parents also prevent them from taking the plunge and taking their future into their own hands, hence their marginalisation.

However, development in the context of globalisation increasingly requires the adoption of another culture. It is Western culture that will give a chance to the competent regardless of gender and age.

Malagasy society, particularly in rural areas, needs to change its trajectory in order to have a new vision and a new perspective within the framework of communal development. This requires community unity and integrity. Young people should prepare themselves and convince their elders to work together to manage the economic situation in the commune.

When a management tool for resources transferred to communities was being developed, young people did not have the opportunity to participate. We refer to young people here as single people who still live with their parents, regardless of age or sex. Perhaps by living with their parents, they are still considered to be children. As a result, they will only be executors while their parents make the decisions. Major decisions are often influenced by the father, even if they are the result of consultation: production methods, major expenditure, family and community affairs. All this explains the passivity of young people, especially when they are confronted with the adult world.

As a result, young people are leaving their land, most of them choosing the *Morondava* region to work as labourers, and some of them are going to try their luck in mining resources such as sapphire in the *Sakaraha* region. This flight of potential young people will weaken the *Mahafale* plateau.

In a nutshell, the assumption of power by elders does not encourage young people to take responsibility.

1.3.2 Forms of empowerment

Before the CoBa management committees were formed, the objectives of the management transfer and other explanations were highlighted at a general assembly meeting in the districts of the communes hosting the management transfers. However, according to the members interviewed, this reference document was being produced at the same time as people were being made aware of the need to set up farmers' groups. This situation was not conducive to the dynamism of the group, which should be pursuing its action in the process of developing a forest resource management tool. The real objectives were not captured or were concealed by temporal sub-objectives, but these were considered to be primordial because the citizen's attention was still monopolised by the creation of these groups.

The case study of the transfer of management of forest resources in the *Mahafale* landscape showed us that not all the inhabitants were interested in the programme because the development of the management tool for these resources took a lot of time and coincided with the growing season. Of the people we surveyed, only 203 out of 800, or 25.37%, had taken part in the process of developing a management tool and in the management of the transferred forests, while 52.37% did not feel very concerned as long as their basic needs had not been met.

People who are not active in the management of transferred forests have not yet assimilated the raison d'être of this management and planning tool, and have confused the management committee with the agents of NGOs working in the field of renewable natural resource conservation and sometimes forest rangers in the landscape. People outside the committee feel more excluded and are hostile to any mobilisation, noting that the projects drawn up by the committee, and especially environmental protection NGOs, are being taken over, in addition to their inability to understand the phenomena taking place in their commune.

The characteristic feature of the management committee is its participation in the participatory diagnosis of the community. However, we noticed from the outset that effective participation is not constant, individual interests take precedence over the interests of the community, and there are gaps in the preparation of projects to be carried out annually because information is not passed on as it should be.

The feeling of belonging to this committee and the cohesion of its members are becoming repulsive, and needs are not being identified or prioritised. The imbalance between hamlets and neighbourhoods is becoming apparent as a result of the non-attendance of some committee members who live in hamlets.

In the long term, the objective of environmental NGOs is to create a project unit or a structure that is representative and close to the population in order to better manage the implementation of forest resource management and conservation activities. However, we have already stated that in order to develop these famous natural resource management tools, the management committee must be created. It plays the role of mediator, influencing and motivating the community to participate actively in the management and conservation of these renewable natural resources.

There are 25 CoBa surrounding the *Tsimanampesotse* park in the *Mahafale* landscape. Most of these CoBa have changed their contract to GCF, whereas before they were under an agar contract. However, some CoBa have chosen to remain under an agar contract. Initially, the information circulated through a specific network: that of the neighbourhood notables and an association set up by one of the neighbourhood notables.

The management committee was therefore set up informally by members of the ruling class, i.e. people close to or belonging to the commune and even to the neighbourhoods: members of the office, local security committee, members of the communal council. The reason for setting it up was to monopolise the power assigned to the committee, so that the members of this ruling class could control the members of the committee and, above all, control everything.

The non-representativeness of the commune's association can be explained by the abundance of active forces working in the commune's development, they argue, but the number of committee members does not exceed 40. We therefore deduce that the choice of member of the management committee explains the way in which the commune or the *Mahafale* region operates. Indeed, during our interviews, we were able to observe that a representative of the *Magnasoa Tane* CoBa reflects the recuperation of power by an entity that wants to become the ruling class in the community, and provokes the rejection of the population, which should be directed towards action.

Malagasy culture, which gives more opportunities to men in all areas, does not allow for the application or even the popularisation of a gender approach that freely allows for the integration of women in development. In the *Mahafale* plateau, this situation is exacerbated by the lack of a women's association or women's group that could mobilise women to take part in development initiatives, whether at community or national level.

As long as there is no equality of opportunity between men and women in our country, community development action continues to be blocked, despite the adoption by the authorities of different development models.

Moreover, the people who are involved in politics are the most politicised and, as a result, they consider themselves to be "*Raiamandreny*", responsible for the population in all neighbourhoods. Taking charge of a development project, in particular the conservation and/or management of forest resources, without being motivated, makes no sense to a population that already submits, in spite of itself, out of fear of politics. Moreover, the population was hardly involved in the process of conserving their renewable natural resources. They were not informed and made no effort to find out about conservation of the forest ecosystem or the actions taken by the CoBa management committee. The absence of any real motivation reduces participation in the long term; the population voluntarily leaves the area, looking for pretexts to avoid their duty.

The population was not really involved in the process of managing their forest resources; they didn't feel represented by the management committee and didn't even know about it. People can recognise the names of the members of the CoBa structure because they live in the neighbourhoods of the commune, but not the existence of this committee.

The socio-organisational capacity of communities can only be developed at one pole. For example, being involved in the development of a forest resource management tool enabled the leaders to develop their capacity to mobilise and group together. The management structure was created, but its weakness was noted once it had been set up. There were around thirty members at the first training session, but this number was reduced during the field diagnostic phase.

To conclude, we can say that the forms of power-grabbing involved in carrying out a project also paralyse most of the inhabitants in their active participation in the protection of these natural resources or in development, especially as a large part of the population seems unable to participate because of their poverty.

1.3.3 NGO dependency

The "top down" approach previously used by development projects and the intervention strategy manifested by a paternalistic State regime have cultivated a spirit of assistance and irresponsibility in the management of public affairs at grassroots level. During previous support programmes, such as the National Agricultural Extension Programme and bank financing under the socialist regime, directives were issued by the authorities and technicians, with the farmers simply carrying out the work. These approaches systematically excluded the population from the decision-making, design and monitoring processes.

In terms of the environment, *Madagascar is one of the seven biodiversity hotspots, with an endemism rate of 80%. It is a haven for the world's ecological heritage, the management and conservation of which are the responsibility of the international community* [GOEDEFROIT S., 2002].

To better manage these assets, donors have adopted the vertical approach applied by the socialist regime. A new implementation structure was therefore set up. Hence the emergence of new players, commonly known as "NGOs", which will implement a development project or a biodiversity conservation project for local or regional development in a southern country like Madagascar.

Moreover, from the point of view of donors, NGOs seem to be well placed to act as two-way agents, through their ability to encourage reciprocal adaptations of visions and strategies. It has to be said that the term "NGO", which has become part of everyday language, is partly misused: in reality it refers to associations with NGO functions, or sometimes de facto NGOs.

The NGO has in fact become a means of economic, social and cultural integration into modernity for intellectuals. In town and country alike, the intellectual elite, who have been to school or university, act as a conduit between the world of projects and the realities of the countryside.

Graduates of Madagascan or, in some cases, foreign universities, the country's intellectual elite have difficulty entering the very limited job market, and suffer from a feeling of marginalisation and devaluation. Finding themselves at an impasse, some returned to the land in their native countryside. Associations of fellow students have sprung up, hoping both to work for the development of their country and to remedy their own situation.

These intellectuals are a supply for international NGOs looking for qualified staff. They make up the bulk of employees, with the exception of management and technical advisory positions.

NGOs operating in Madagascar fall into two categories: international and national.

International NGOs, i.e. organisations with headquarters abroad, participate alongside donors and the State in the design of sustainable development strategies, such as nature protection NGOs (WWF and Conservation International) and NGOs or projects that function as agents for implementing the objectives pursued by donors, such as LDI (Landscape and Development Intervention).

These structures have very substantial financial resources at their disposal, provided by the World Bank and other donors and, depending on the case, also benefit from contributions from Western donors. They work in fairly large areas. They always work over a long period of time, sometimes more than a decade.

Malagasy national NGOs, which are on a smaller scale, have been able to forge privileged links with Western players, thanks to the seriousness they demonstrate and their political support. These structures are often funded by other international NGOs. They have been sidelined from discussions on the environmental strategies to be implemented, and essentially play an executive role. Operating in one or more municipalities as part of projects that are generally multi-year, they carry out the same tasks as their predecessors, but with more limited resources.

Finally, at the lower end of the scale, many small Madagascan associations work as subcontractors for higher-level structures. They work in specific geographical areas. Their missions are brief: feasibility or impact studies, socio-economic diagnoses, training in various fields, etc.

NGOs are financially dependent on donors and cannot assert themselves because they are under the shadow and orders of donors. It is the donors who dictate the methods and strategies to be adopted.

In her book entitled "Des associations des villes aux associations des champs en pays betsileo", Sophie Moreau noted that:

> *"This financial dependence leads them to model their interventions on the expectations of the donors who commission the environmental policy, at the risk of disconnecting them from farmers' realities. This certainly does not encourage them to be innovative or critical. NGO workers rally to global environmental standards, either because they actually adhere to them, or because they hide their disagreement in order to protect their own interests".*

In general, NGOs take no account of the calendars or traditions of Malagasy farming societies. They follow the donor's calendar. It is difficult for the rhythm of technical and financial partners to coincide with the farmers' calendar and not generate the continuity that is essential for the adoption of innovations. Finally, the lack of attention paid to the grassroots community is due to the ambivalent attitude of NGO staff towards it.

For city dwellers, the farming world represents their identity roots, as they are often descendants of farmers. But through their studies, they have done everything to distance themselves from it. They rarely visit their home village and have virtually no experience of farm work. So, in the city, the founding values of Malagasy identity such as *fihavanana* are in decline.

The reluctance of peasants to embrace all that the city has to offer leads to contempt for the countryside on the part of city dwellers. So working for an NGO brings them closer to Western models. Between donors and farmers, they are more inclined to adopt the values of the former than those of the latter.

Field workers, subject to the constraints of life and work in the bush, often have only one thought in mind: to return to the city. NGO workers are the first to be confronted with the semi-failures of their actions, which they often attribute to the unwillingness or laziness of the farmers. As a result, they are not always the best placed to make the most of farmers' representations.

In conclusion, the dependence of NGOs, whether international or national, means that they are fragile when it comes to carrying out development or conservation activities. They do not have the choice of whether or not to adopt a strategy and/or approach to carrying out their activities.

It is the donors who dictate what they should do and what their objectives should be. As a result, they are instrumentalised by these donors. They are accountable only to the donors. In a word, they are accomplices and thus participate in *"ecological interference"* [ROSSI G., 2000] or even *"green neo-colonialism"* [HUFTY M. & RAZANAMANATSOA A. et CHOLLET M., 1995] committed by hyper-conservationist donors.

To conclude this chapter, the *Mahafale* plateau lies in the south-west region, bounded to the north by the *Onilahy* river, to the south by the *Menarandra* river, to the east by the RN10 and to the west by the Mozambique canal. It has a sub-arid climate. It is a region of plateaux and plains that forms part of the Saharan regions. It has a long dry season lasting 7 to 9 months. The rainy season is often very irregular and always poor in rainfall. An area under constant threat from locusts, it is inhabited mainly by the *Mahafale* inland and the *Tanalana* on the coast.

As for social organisation, the *Mahafale* social fabric has been torn apart. The community is frustrated by external domination (central *Fanjakana*, evangelisation with cultural domination, development NGOs and neo-colonialists or donors) but also internal domination by the educated. These forms of domination lead to a general crisis in the villages. This crisis manifests itself in various forms of protest, such as withdrawal, possession and rural exodus.

The *Mahafale* communities, weakened by both internal and external domination, are therefore being encouraged to conserve and manage their natural resources. In the next chapter, we will try to identify the issues surrounding the conservation of natural resources.

Chapter Two : Conservation of natural resources

We are going to try and explain what conservation or natural resource management means, using the protected areas system in Madagascar as a starting point.

We will first define what is meant by the transfer of natural resource management, followed by a process for implementing this transfer of management and a social and environmental protection plan.

2.1 The historical foundations of Madagascar's conservation policies

2.1.1 The Madagascar Protected Areas System (SAPM)

Madagascar's environmental policy was adopted in the 1990 Environment Charter. A National Environmental Action Plan (PNAE) has been drawn up. This is divided into three 5-year Environmental Programmes[58] (EP). The implementation of these environmental programmes has seen the creation of two generations of protected areas: the first generation concerns the network of national parks managed by MNP (National Park, Integral Nature Reserve and Special Reserve). The second generation is made up of the new PAs created under the aegis of the Direction Générale des Eaux et Forêts (Natural Park, Natural Monument, Natural Resource Reserve and Harmonious Protected Landscapes).

These two generations of Protected Areas fall into the different categories of PA defined by the IUCN and are destined to form the Madagascar Protected Areas System (SAPM).

In addition, the commitment made by Madagascar's government leaders at the Durban World Conference in 2003 to increase the surface area of protected areas is a key objective of Madagascar's environmental policy for the period 2007 to 2012. The aim is to increase the surface area of protected areas to ensure the conservation and enhancement of the country's terrestrial, lacustrine, marine and coastal biodiversity.

A number of interministerial decrees have made it possible to place large areas of natural spaces under temporary global protection[59] and in definitive decrees to arrive at the current 122 protected areas[60] .

In setting up this system, SAPM has two objectives:

- conserve Madagascar's unique biodiversity (ecosystems, species, genetic diversity);
- maintain ecological services and the sustainable use of natural resources for poverty reduction and sustainable development.

In the legal-institutional framework, the state laws[61] of 2008 and the Code of Protected Areas (COAP) of 2015 define the legal regime for public protected areas.

---

[58] PE I from 1991 to 1995, PEII from 1996 to 2000 and PEIII from 2001 to 2005.

[59] Interministerial orders 52005/2010 of 20 December 2010 and 9874/2013 of 05 May 2013.

[60] Decrees of 21 April and 28 April 2015 in Annex 2.

[61] These laws are supplemented by a specific regime announced in Law 2005-019 of 17 October 2005 (art. 38), a law setting out the principles governing land status.

After the Rio Summit in 1992 and the Durban Congress in 2003, the conservation of forest resources and the creation of marine protected areas are now major international issues.

In Madagascar, to build up a respectable image, the leaders used this to extend the protected areas, particularly marine protected areas, which mobilise donors, international NGOs and State institutions.

2.1.1.1 Forest conservation policy

Forest management in Madagascar is evolving, regardless of the regime and institutional innovation applied (monarchical regime; democratic regime).

Measures and rules have been adopted to ensure the efficient management of these natural resources. The leaders of these different regimes are therefore aware of the importance of natural resources, particularly forests. This is why they are considered as natural capital. These measures and regulations vary according to the time and place in which the natural resources are located.

2.1.1.2 The origins of Madagascar's forestry regulations

In Madagascar, local management of renewable resources has deep historical roots. As early as the beginning of the 19eme century, King *Andrianampoinimerina* [1787-1810] realised the importance of the Malagasy forest and considered these resources to be part of his heritage. He banned bush fires as soon as the fire season arrived, and he also did not authorise the manufacture of charcoal for blacksmithing. However, he let the Malagasy obtain these forest resources for themselves.

Since then, the Malagasy people have avoided the complete and irremediable disappearance of the forest. This customary prohibition provided an initial framework for the management of Madagascar's forests.

According to Sourdat in 1998, *Andrianampoinimerina*'s policy was therefore based on :

(i) deterrence through threats;

(ii) participative management;

(iii) raising awareness, showing that the forest is ultimately the last resort of the poor.

In 1881, Queen *Ranavalona* II published a new code known as the "Code of 305 Articles". This code was promulgated on 29 March 1881. It contains 6 articles relating to forests [101 to 106] out of a total of 305. The Queen provided a formal framework for forest management and specified the penalties for offenders. Article 101 states: "*The forests must not be burnt; those who burn them will be put in irons for 10 years*". Nearly 25 years before French colonisation, the State was demonstrating its interest in Madagascar's renewable natural resources.

At the beginning of the colonial period, the rules laid down by the Malagasy authorities did not protect the forest ecosystems that were expected. However, biodiversity was unanimously recognised as exceptional but threatened because forest policy management at that time was in the hands of the colonisers. As a result, the environment widely described deteriorated throughout the 20th century. Assessments of this degradation are still debatable.

The rise of colonisation in Madagascar led to the introduction of new rules imposed by the colonisers.

From 1896 to 1913: First legislative and regulatory texts; The fundamental axes of forestry policy in countries under French domination were found in the grounds of the forestry code of 21 May 1827. It would appear that the aim was to transpose the objectives of metropolitan forestry legislation.

So soon after the colonial conquest in January 1896, a forestry service was organised, but the job of head of the forestry service was quickly recognised as unnecessary and forest management was placed under the orders of the administrator in charge of the colonisation office. Governor General Gallieni himself did not seem convinced of the importance of this sector.

From 1896 onwards, the colony drew up a number of texts relating to forest exploitation, land clearing, user rights and bush fires.

In addition to Gallieni's questioning of the very principle of the existence of a forestry service, the succession of texts that were to govern the sectors until the end of the 1920s were inspired by French and Algerian texts.

Far removed from the realities of Madagascar, those in charge are aiming to manage the use of timber for commercial purposes, limiting the rights of local people to use it and banning "*Hatsake*" or slash-and-burn cultivation.

1930: A framework decree creating the world's first ecological forest reserve was drawn up in response to the ineffectiveness of previous legislation. It was noted that the Malagasy habit of destroying forest stands by fire continued to be tolerated by the local authorities; this tolerance was sometimes the price of accurate tax collection, submission to recruitment or unpopular health measures, and that administrative tranquillity was obtained at the expense of the forest. The forest was the object of administrative blackmail: social peace in exchange for authorisation to burn. What's more, the colonial need for agricultural land also meant that forest areas had to be cleared.

At the end of the 1920s, the colonial government noted that there was large-scale clearing[62] of the forest, so it drew up and promulgated the decree of 25 January 1930, the introductory report announcing that :

> *"One brutal fact dominates the forestry issue in Madagascar: at least 100,000 hectares of forest disappear every year out of a total surface area estimated at 10 million hectares".*

This decree refers only to local contexts, customs and other traditional management methods, and only by reference to the 305-article code (protection against fires; clearing and use of wood), which gives these texts a clear orientation towards strict protection of the forests. This text became, for several decades, the legal framework for the forestry regime. Thirty years later, the forestry services once again became independent of the agricultural services through the creation of six forestry districts. This decree reorganised the forestry system in Madagascar and stated in Article 1 that *"Forests belonging to the State, the colony, the communes and other public establishments are inalienable and imprescriptible".*

---

[62] Once they had the land they needed for their farming activities, the colonists decreed to reorganise Madagascar's forestry regime in their favour.

A change occurred in the 1990s, in a context of liberalisation and democratisation. This change was reflected in the adoption of a policy of biodiversity conservation, which led to the introduction of territorial mechanisms. Centralised strategies, deemed ineffective, were replaced by participatory approaches. New institutional rules are accompanying decentralisation and the involvement of local people in resource management. The changes are therefore marked by the inclusion of environmental protection at local level, with the transfer of management responsibility to local communities.

2.1.1.3 The evolution of these regulations over time

Madagascar is renowned worldwide for its exotic nature and exceptional biodiversity, with one of the highest rates of endemism in the world. Nevertheless, the environment has suffered very rapid and catastrophic degradation due to heavy deforestation and worrying erosion, the result of inappropriate and degrading farming and pastoral practices. In addition, demographic growth is leading to land clearance, which in turn leads to deforestation, increased climatic variability and, above all, worrying rural poverty.

At the beginning of the 20th century, during the French colonial period, Alfred Grandidier [Grandidier, 1928], a tireless walker, estimated that 200,000 hectares of the Grande Ile were deforested every year as a result of fires and clearing [Bertrand and Randrianaivo, 2003].

On the other hand, the 1995 report on the environment in Madagascar states that *"according to estimates based on old aerial photographs from 1950, forests and laws covered 14 to 16 million hectares representing 24 to 28% of the national territory. Forest cover is currently estimated at 12 million hectares and natural forest today covers only 9 to 10 million hectares, or 16 to 17% of the island"*[63] .

It is therefore difficult to seriously and accurately quantify the reduction in forest areas in Madagascar.

However, recent studies have provided some interesting and again contradictory details. The annual rate of deforestation is estimated at 0.83% for the period 1990-2000. It fell by 0.53% per year between 2000 and 2005[64] . This report estimates that 9.4 million hectares of natural forest remained in 2005 and that more than 1.2 million hectares had been lost between 1990 and 2005 [fifteen years].

For over a century, Grandidier's "estimates" have generally been regarded as benchmarks and included in official documents. Every year, certain organisations or authors repeat these assessments, or others that are even more catastrophic, in order to justify to politicians and decision-makers a policy that further excludes populations from access to forest resources. Others use questionable evidence to overestimate the extent of deforestation.

The aim of repeating Grandidier's estimate (1928) is clear: to legitimise the recurrent accusations of environmental degradation, bush fires and deforestation levelled against rural populations in order to justify repression and the exclusion of the latter from natural areas, but above all to put pressure on the Malagasy state to adopt Abel Parrot's 1925 proposal that Madagascar's forests should be managed by local authorities and to introduce the French communal forest regime into Madagascar. Fifty years ago, this rhetoric worked to the advantage

---

[63] ONE et al, 1995.

[64] MEFT, USAID and CI, 2009

of foresters [Lavauden, 1934], but for the last two decades it has been used to the advantage of environmentalists. Through these discourses, it is a recurring conflict over control of and access to biodiversity resources that needs to be deciphered.

You can't change regulations if you're convinced that the current ones have reached an impasse.

If we go back to the local management of natural resources under the Malagasy royalty, several authors point out that it made the forest its inalienable domain but, at the same time, left the local populations a great deal of latitude to manage this domain directly, as there were few prohibitions.

It was during colonisation that the policy of repression and expulsion of populations was applied. It all began with the creation of the forestry service in 1896. As early as 1897, a circular from the Governor General prohibited pasture fires and a decree regulated the exploitation of forest products. *Tavy* and *hatsake* were banned by decree in 1900. After 1904, a short period of less repression began until 1907. From 1907 to 1917, a succession of decrees strengthened the repressive arsenal and the forestry service was expanded until 1930, when the 1930 decree drafted by Lavauden introduced performance bonuses for forestry officers based on the number of official reports drawn up.

The colonial policy was justified by the concern to preserve the forests from the irreparable destruction that threatened them [Lavauden, 1931], but at the same time the areas granted for timber exploitation increased between 1897 and 1901 from 700 to 101,630 ha, 43% of which concerned the Manjakandriana forest alone [Bertrand, 2004]. This policy was clearly intended to be repressive. But by 1937, it was clear that this repressive policy was leading to a dead end [Coudreau, 1937][65] .

Colonisation considerably reduced Madagascar's international horizons, with relations, both commercial and cultural, centred mainly on the French mainland. Once independence was restored, successive leaders chose to establish diplomatic relations with strategic partners that corresponded to the ideology of the time and their needs.

Under the First Republic, the regime led by President *TSIRANANA* chose, for example, to maintain a French-style administration and mode of governance. Nevertheless, it opened the doors of Malagasy diplomacy to countries such as West Germany, the United States and South Africa. In the aftermath of the 1972 crisis, links were forged between Madagascar and the USSR, China and North Korea. From 1972 to 1974, President *Ratsiraka*'s regime attempted to break with France, an initiative that was reinforced when he came to power in 1975.

It is normal for cultural ties to unite two countries that have been "intertwined" for almost 70 years. On the other hand, it is not natural for the former coloniser to continue to rule in the shadows the country to which he is supposed to have restored freedom. It is even less normal for it to continue to manipulate its former property on the world stage in order to serve its own interests. This phenomenon has a name: Françafrique[66] .

---

[65] Bertrand A., Rabesahala N. H. and Montagne P., 2009

[66] Françafrique is a kind of tacit pact between the French Republic and its former colonies. It guarantees African leaders continued protection from the vagaries that can hamper their power, asylum in the event of prosecution, and certain administrative, property and financial facilities that ordinary people have no idea about. All this in return for economic facilities, secret financing of political parties and investment priorities in the countries concerned.

Because of the political choices made by successive leaders, but above all because of globalisation, Madagascar has many partners. As a result, regional and international powers are involved in games of rivalry, cooperation and alliances.

The political strategy or "realpolitik"[67] of international powers mobilises "hard power", in particular military and political power; "soft power", whether religious, linguistic, media or cultural, develops diplomacy of influence. The strategies of the main players are geared towards conquest, presence in order to seize opportunities, but also protection against nuisance or the risk of domino effects. They combine structural powers (security, production, finance and knowledge) and relational powers (Strange). Spaces structured by networks differ from the official borders created by the violence of history, even if they paradoxically confirm the territorial basis of states while helping to weaken them. Some powers, such as France and China, have a high profile, while Israel, the Gulf States and Iran operate through intelligence services, particularly military, the religious sphere for Sunni and Shiite powers, and networks (Lebanese, Indian diasporas).

In short, Françafrique is a great story of hypocrisy that is not about to come to an end, because the historical ties that bind Madagascar to France, ties that the Malagasy people did not choose, continue to weigh heavily on their destiny.

And some authors are already talking about Indeafrique and Chinafrique. Apart from non-repayable aid, the latter use another strategy to work or cooperate with African countries. They have used direct manoeuvres such as land purchases. The map below shows that demand exists and that Madagascar has potential. Madagascar ranks as one of the 9è top land-selling countries in the world.

[67] Political strategy whose sole aim is efficiency or political power based on force.

Map 7: Map showing the world's biggest sellers and buyers of land

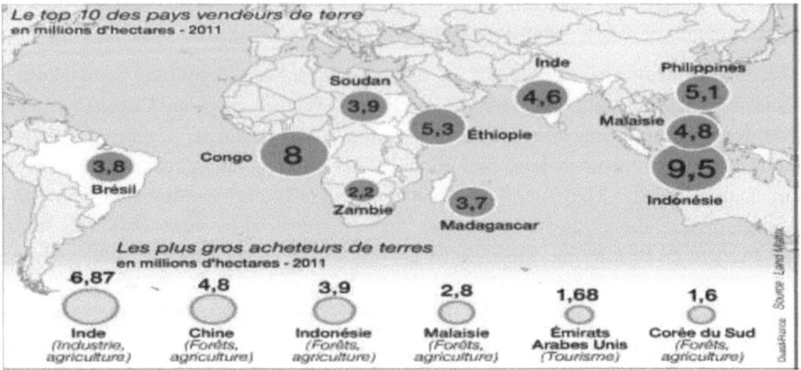

Source: Land Matrix, 2012

Faced with the threat posed by Indafrica and Chinafrica, France has defined 17 poor countries as priorities for aid[68] . Economic diplomacy, which must be translated into return effects, leverage and influence, complements both a policy aimed at transparency and human rights and military interventions. Development cooperation and cultural cooperation, although strategic in the long term, have lost weight in the face of environmental and military priorities.

In the case of Madagascar, the promulgation of the Environment Charter in 1990 was the starting point for a change in the State's approach to the management of renewable natural resources, which until then had consisted of giving the administration exclusive responsibility for this management. The 1990s were a period of socio-political upheaval in the country, following François MITTERRAND's speech [June 1990] at La Baule on the adoption by African countries of a political system based on democracy, culminating in the adoption of a general policy based on liberalisation of the economy, disengagement of the State from the productive sector and decentralisation. As regards the management of natural resources, it was noted that, due to a lack of resources of all kinds, the administrations in charge of this activity were increasingly powerless to stem the tide of destruction and illegal exploitation of these national assets, due to the existence of open access situations.

The second observation is that people living close to resources are both victims and, to a large extent, responsible for the spiral of degradation of Madagascar's natural environment.

The third observation is that village communities in several parts of the island apply certain forms of traditional resource management that are not valued because of the system of exclusivity applied by the authorities in this area.

As a result, France has latched on to the environment through French cooperation and the Fonds Français pour l'Environnement dans le Monde (FFEM). Traditional donors are using legal instruments to stem the Chinese and Indian "invasion" of Africa, including Madagascar.

---

[68] Hugon P., March 2016. Most of these countries are least developed countries in sub-Saharan Africa, including Madagascar.

The first instrument to be developed was a legal one, Law 96 025 of 30/09/96, known as the Gélose Law, which was passed and promulgated just eighteen months after *Antsirabe*; it concerns all renewable natural resources, forestry, continental and marine fisheries and pastoral resources; it sets out the institutions to be involved and the rules and conditions for developing and implementing the transfer of natural resource management; this mother law will be gradually fleshed out by implementing regulations.

The forestry sector was evolving to define the broad outlines of its new forestry policy, drawing in particular on the experiences of participative forest management in the *Menabe* region, supported by Swiss cooperation.

Thus, the law revising forestry legislation, which was promulgated on 08/08/97 under number 97-017, and its application text in terms of exploitation under number 98-782 of 16/09/98, referred to the Gélose law in terms of forest resource management transfer contracts until 2001. In that year, the forestry administration issued decree 2001-122 of 14/02/0, specifically for the forest resource management transfer contract known as Gestion Contractualisée des Forêts (GCF); this decree is an alternative to the Gélose contract, which is considered too complicated due to the use of a mediator, the obligatory signature of the Mayor of the CoBa's Commune, and the relative security of land tenure.

2.1.1.4 Marine biodiversity conservation policy

Compared with the policy of conserving forest resources, the policy of protecting marine biodiversity has always been somewhat vague. The law on the sea or on the conservation of marine biodiversity was not one of the laws promulgated during the feudal period. It is true that *Andrianampoinimerina* said during his reign that *"ny ranomasina no valam-parihiko"* literally it is the sea that delimits my territory. The *Merina* kings and queens who succeeded each other "recognised by Europeans as kings or queens of Madagascar" were not concerned only with their territory: central *Imerina*.

During the colonial period, marine biodiversity conservation policy was always unclear, as the Malagasy sea was governed by French law.

It was not until the 1990s, when the Malagasy government adopted the Environment Charter, that Madagascar opted for a biodiversity conservation policy. This policy was reinforced by Law 2001-005 of 11 February 2003 on the Protected Areas Management Code [COAP].

2.1.1.4.1 Setting up marine protected areas (MPAs)

According to the usual IUCN definition, Marine Protected Areas only concern the marine environment[69]. We include in this category marine and coastal protected areas (MCPAs), which include both marine and terrestrial components.

The Marine Protected Area is an area of management and governance that often functions as a closed system, with minimal links to the catchment areas and the local socio-economic environment, other than to minimise poaching or generate revenue. From a tourism point of view, the MPA is an attractive area, sometimes generating a concentration of hotels and scuba

[69] "Any intertidal or subtidal area and its underlying waters, flora, fauna, and historical and cultural resources, which have been set aside by law or other effective means to protect all or part of the area so delimited".

diving clubs along its edges, whereas for fishermen, the MPA is a deterrent (shifting of fishing effort to other areas and other species), but also an attractive area (edge effect).

Madagascar's first marine protected area was officially created in 1989 with the *Nosy Antafana* Marine Park, part of the *Mananara* Nord Biosphere Reserve, followed by the three marine parcels of *Masoala* National Park. Both are located at the tip of Antongil Bay. The total surface area of these two marine areas to the east is less than 100 km2.

In 2003, at the World Parks Congress organised in Durban by the IUCN, the objective of classifying 20% of the world's marine waters as MPAs within 20 to 30 years was adopted. Island states are heavily involved in this process [CHABOUD et *al.,* 2008]. In Madagascar, the current regime is committed to tripling the number of protected areas, particularly in forest and marine environments. Between 2003 and 2009, the surface area dedicated to the protection of biodiversity should increase from 1.7 million to 6 million hectares, including 1 million hectares in the marine environment [DAVID G. et *Al.*, June 2008].

2.1.1.4.2 Institutional context of MPAs in Madagascar

Since 1990, ANGAP (today's MNP) has directly managed protected areas. It is also responsible for promoting and managing ecotourism in MPAs. The MNP believes that the success of protected areas depends on the support of local populations, and that this support will only be achieved if biodiversity conservation is accompanied by development operations that benefit these populations. The Gélose law of 1996 provides for a co-management contract between the State, through the intermediary of the Commune, and the neighbouring grassroots communities of natural resources. This process is supported by the SAGE association, which works in the following areas respectively

- participatory planning, with the promotion of consultation platforms involving the various categories of regional and local players;
- management transfer ;
- capacity-building and social communication to enable local communities to take charge of the sustainable management of their resources.

If we refer to the evaluation reports on the first agar contract for the CoBa on the *Mahafale* plateau, the SAGE association's mission was doomed to failure. First of all, the CoBa were set up hastily. Perhaps it was a first experience. There was also a problem of communication or persuasion, as the principle of management transfer was unclear to the communities during our interview with the board members of the CoBa supported by SAGE but not renewed by the NGOs working in the area.

Until now, the creation of marine protected areas has always been the responsibility of the State. However, the State may no longer have the means to fulfil its prerogatives as creator and manager of marine protected areas, because it is still difficult to set up these MPAs. The financial and social cost of conservation, already high for terrestrial protected areas, remains an unresolved problem for future MPAs.

Donors have taken control of the process of creating MPAs (by controlling the financing and running of the MPA, which government departments are unable to do). They are not free of the need for the State, which is essential for ratifying the legal status of the MPA, zoning, policing, etc. The donors are constantly approaching the State services. Once the legal framework is in place, the landlords often offer to manage it or sponsor its delegation to an NGO or agency. In this way, the State has left and continues to leave the field open to the action of numerous international NGOs or public utility associations.

Madagascar's reputation for biodiversity is leading international NGOs to become increasingly involved in the conservation of this biodiversity, particularly in the creation of protected areas. This consequently reduces the State's effective capacity to intervene, even if the NGOs cannot bypass the State on all the administrative aspects and the formalisation of the MPA.

The management of MPAs in Madagascar by international NGOs is very delicate because they impose their concepts and their point of view, and define the problems, the objectives and the means of action. If Madagascan public policy is therefore an internationally inspired policy, we may well wonder about the degree of freedom the country has in defining its development choices and public action.

In trying to answer this question, Ranaivomanana et *al.* affirmed that in *Madagascar in particular, on the one hand, there is an inordinate eagerness on the part of the State to achieve colossal objectives (so as not to be out of step with the global movement) with almost insignificant resources, and on the other hand, there are the gesticulations of NGOs, with a strong ecological label, who are piloting the establishment of MPAs; this insertion is further confirmed by the perception of the local population: MPAs, with their prohibitions and restrictions, always represent a constraint, a loss of income and an obstacle to their daily lives* [Ranaivomanana et al, 2010].

It is clear that the creation of marine protected areas has complex effects on fisheries, such as exclusion from protected areas and the reallocation of fishermen in terms of space and target species. It also implies a repositioning of fisheries within coastal activity systems, with direct exploitation of marine resources being partially or totally ruled out within MPAs in favour of tourism or non-use activities.

However, the people living along these coastlines, the majority of whom only fish for a living, risk having to put up with this system. Fishermen are therefore often the main opponents to the creation of marine protected areas, or at least their perceptions of the latter are more negative than the average for the coastal population, given that the effects (positive or otherwise) of the setting aside of marine reserves on marine populations are only perceptible after a minimum of five years.

The sustainability and social acceptability of these guidelines decreed at international level also give rise to a number of considerations [Chaboud, 2007].

For example, the temporary octopus reserves in the *Velondriaka and Andavadoaka* marine protected areas are no longer respected by local people. They fish at night during the closed season. They are no longer governed by established conventions. Production in these reserves has been falling year on year since 2014.

2.1.1.4.3 *Nosy Ve* Marine Protected Area - *Androka*

The Nosy *Ve-Androka* marine protected area is located in the south-west region, in three rural communes, namely *Beheloke* in the District of *Toliara* II, *Itampolo* and *Androka* in the District of *Ampanihy*. The *Nosy Ve-Androka* MPA is a category V cluster protected area. It covers a surface area of 92,080 hectares, including 28,829 hectares of hard cores and 63,260 hectares of buffer zones. It has eight (8) category II hard cores. It is part of Madagascar's network of National Parks and Reserves.

To deal with the disruption caused by the restriction of coastal communities, they hope to have a development programme:

- drinking water supply ;
- the construction of social infrastructure such as schools, CSBs, roads, markets, etc., especially the road linking *Androka and Ampanihy*, which will make it possible;
- the desilting of *Ampalaza* Bay;
- the eradication of *"raketa mena"*;
- combating the advance of the dunes ;
- and setting up a central purchasing unit.

It should be remembered that the majority of coastal inhabitants are *Tanalana* and *Vezo*. The *Tanalana* suggest desilting the bay so that fishermen (*Vezo* or not) can continue their fishing activities. To increase their cultivation areas, the communities suggest combating the advancing dunes and eradicating the *raketa mena*. The central purchasing office is intended to supply agricultural inputs. This means that the *Tanalana* are aware that fishing grounds will diminish and they are ready to devote themselves entirely to agricultural activities.

However, the project promoter and technicians have found no other way to compensate the fishermen[70] than by establishing exclusive fishing zones, providing fishing equipment (nets, canoes, etc.) and training in improved fishing techniques.

As far as the technicians are concerned, fishermen will only benefit from this compensation if they form an association.

Since the project began in 2012, the community programmes included in the social and environmental protection plan have not been implemented. The same goes for those proposed by the technicians and the project promoter, as the cost of this compensation is estimated at 47,000,000 Ariary.

The demarcation of core and buffer zones, the creation of *dina* to manage this protected area and the creation of management committees for each site have already been completed. However, there has been little training and support for the management committees due to a lack of funding. As a result, the management committees cannot implement the activities defined in their management tools.

Apart from funding, the MNP has specialised in the terrestrial environment and is finding it difficult to manage the marine environment. It is still in an experimental phase. This is why there has been a delay in implementing activities to protect the *Nosy Ve-Androka* marine biodiversity.

But why haven't the *Tanalana "Toko bey Telo"* reacted to the creation of these *Nosy Ve-Androka* protected areas?

[70] 2757 people are affected by the project and 382 households are affected. They are vulnerable. 378 households are major Project Affected Populations (PAPs) and 4 are minor PAPs, according to the 2009 census.

In the past, the fishing activity along the coast was occupied by the *Vezo*. They immigrated to the southern coast, which was already occupied by the *Tanalana*. The *Vezo* found that the area was still under-exploited, so they settled there. For the *Tanalana,* the *Nosy Ve-Androka* MPA is a promising way of recovering and/or regaining control over the management of their coastal and marine territories.

Because of a long period of drought affecting the *Mahafale* plateau, their farming activities have deteriorated. They believe that their land has become barren because of turtle blood, even though they used to practise the *"tsotse"* or annual sacrifice of the most beautiful zebu to ask for rain. The *Tanalana*'s solution was to replace their farming activities with sea fishing. What's more, they wanted to be masters of their land. They are the ones who run the management committees for the marine protected areas and the *Vezo*, who are considered to be the "major population affected by the project" or PAPs, remain outside the committee.

After the establishment of the *Nosy Ve-Androka* MPA, to escape this new organisation, the *Vezo* migrated to *Belo* sur Mer or the Barren Islands[71] in *Maintirano*. They stayed there for 8 months before returning to the southern coast.

2.1.2 The origins of conservation in Madagascar

2.1.2.1 Conservation of forest resources

Forest resources have been severely degraded as a result of human activity. Following the plundering of forest resources during the colonial period, population growth has been very rapid, especially in urban areas. This means that the population's socio-economic needs are increasing. As a result, in order to meet these demands, natural resources are being over-exploited, particularly forest resources, such as wood for building, firewood for domestic energy, etc.

In addition, the poverty currently experienced by a large proportion of the population has as an immediate consequence the search for additional income through increased exploitation of resources, which is often anarchic and even illicit. The rural economy in general is still based on the maintenance of extensive agricultural practices, for example, the traditional practice of subsistence farming on slash and burn or "*hatsake*" by a growing number of farmers, while the area of natural forests has shrunk, with no possibility of regeneration.

In order to limit this degradation, the State has developed various measures, through the application of texts that are the basis of public action in the forestry sector, relating to the management and exploitation of forest resources[72] . There has been a succession of such texts, but more recently the basic texts have been supplemented by special provisions relating to conservation in the context of environmental concerns[73] . However, when these texts and laws are adopted, there are problems with their application, as they are essentially based on a system of prohibitions and restrictions on use, accompanied by penalties, which are all the more ineffective because the resources available to the forestry administration are very inadequate

---

[71] The Marine Protected Areas in the area are currently being set up, and the Vezo migrants are opposed to this MPA project.

[72] Governed by the decree of 25 January 1930, order 60-127 on land clearing and order 60-128 of October 1960 on vegetation fires.

[73] Law no. 90-033 of 21 December 1996 on the environment charter.

in relation to the scale of the pressure being exerted on forest resources. For example, the control of bush fires, logging and land clearing.

The development of the informal sector is another manifestation of the forestry administration's loss of control over the forest products sector.

One of the reasons for the degradation of forest resources is the lack of responsibility shown by those involved. As a result, they exploit the forest as they see fit. Moreover, forest products play an important role in Madagascar's economy. However, this economic potential is not sufficiently exploited. Such a worrying assessment of Madagascar's forestry situation calls for a redefinition of the country's forestry policy, in which the forest's assets must be exploited rationally.

2.1.2.2 The fundamental principles of forestry policy

Madagascar's forestry policy is based on six basic principles that serve both as permanent criteria for guiding action and as benchmarks for assessing results. These principles are part of the long-term perspective that characterises the management of forest resources. They refer not only to the assessment of the country's forestry situation and the resulting findings, but also to the many services that the forest provides to society through its ecological, economic and social functions.

The first principle is that forestry policy should be consistent with national development policy. National forestry policy should be as closely aligned as possible with the country's overall development policy. For this reason, conformity with the objectives of policies adopted in the field of rural development and the environment is a basic criterion for the formulation and implementation of forestry policy.

Similarly, the forestry strategy must fit in with the national options for decentralisation, disengagement of the State from the productive sector and economic liberalisation: a strategy implemented as part of the structural adjustment programme developed by the traditional donors, namely the International Monetary Fund (IMF) and the World Bank (WB).

The second principle emphasises the conservation of forest resources through appropriate sustainable management, since these forest resources are renewable but not inexhaustible. It is necessary to guarantee the best possible conditions for their conservation and renewal, for the benefit of the nation and future generations. This means improving management methods rather than reinforcing prohibition systems. Sustainable and appropriate management methods should make it possible to reconcile the satisfaction of economic and social needs, on the one hand, and the preservation of an overall balance between extraction and reconstruction of the resource, on the other.

The third principle concerns limiting ecological risks. Forest cover plays an important role in preserving certain balances, the disruption of which can cause major and sometimes irreversible ecological problems. These problems cause considerable direct or indirect social and economic costs for the community. Public authorities must act to prevent such problems arising in the medium or long term. The adoption of measures aimed at protecting or restoring forest cover in at-risk areas is part of this approach.

The fourth principle is based on the forestry sector's contribution to economic development: forests form the basis of a sector of economic activity designed to meet the domestic needs of households and supply markets with forestry products. The operation of this sector can be significantly improved in terms of harvesting, processing and marketing, in order to increase its contribution to national economic development. This contribution should be reflected in satisfying the growing demand for better quality products, creating more jobs and increasing the forestry sector's share of national income. Forestry development, as a source of income, can and must be conducted in such a way as to ensure that the capital represented by the forest is maintained and even increased in the long term.

The fifth principle concerns the empowerment of local stakeholders in the management of forest resources. The sustainable management of forest resources cannot be envisaged without the involvement of the various local players concerned. The national forestry strategy will therefore give priority to involving them in the management of forest resources by giving them responsibilities within the framework of management contracts.

The sixth and final principle reflects the adaptation of forestry actions to the realities of the country. The effectiveness of forestry actions must be sought above all in their adaptation. This is an important criterion for prioritising the actions to be taken. This means, on the one hand, focusing actions on the most crucial forestry issues arising in the various ecological regions and, on the other hand, giving priority to actions whose implementation is compatible with the technical, economic and organisational capacities of the players concerned. Finally, given the inadequacy of the financial and human resources likely to be allocated to the implementation of forestry policy, their optimal use must be sought with reference to the notion of cost-benefit.

2.1.2.3 Forestry policy objectives

Madagascar's forestry policy is based on four main guidelines. These guidelines determine the areas of intervention to be prioritised in order to remedy the main problems set out in the assessment of the forestry situation, while referring to the fundamental principles defined above. They consist of halting the process of forest degradation which, given its scale, constitutes a threat to the sustainability of the forest and biological heritage. Indeed, if the current rate of degradation continues, the forest heritage is threatened with extinction in the relatively near future. It is therefore imperative to do everything possible to halt this process. The corresponding objectives will therefore focus on the conservation of biodiversity and forest ecosystems. They will consist of :

- support alternative rural practices ;
- help control bush fires ;
- preserve the forest heritage and the major economic balances.

2.2 The concept of transferring the management of natural resources

Despite independence, it was not until the late 1980s that Madagascar's repressive and exclusive forestry policy was called into question by the proposals of the GPF project (Forest Management and Protection, financed by the World Bank), and by the initial guidelines of the Environmental Action Plan (PAE), the preparation of which in 1986 initiated a major reorientation of its environmental policy.

As early as 1989, it was proposed to break with the policy of exclusion and to begin the process of reorienting forestry policy and reforming forestry regulations and taxation, which was to be completed in 1997. Article 6 of Law 90-033 of 21 December 1990, supplemented by Law 97-012 of 6 June 1997 on the Environment Charter, states in its last paragraph: "to help resolve land tenure problems".

As part of the GPF project, it was proposed back in 1989 to make the forest a means of economic accumulation and development of forestry activities, or to make trees and forests profitable investments without having to destroy these exhaustible resources. The essential objective of this policy is to reconcile the population with its environment with a view to sustainable development.

Various approaches have been adopted to achieve this objective, including participatory management of renewable natural resources, now known as Gestion Locale Sécurisée (Gélose) under Law 96-025, and participatory forest management, now known as Gestion Contractualisée des Forêts (GCF) under Decree 2001-122.

This sustainable management of renewable natural resources cannot, of course, be envisaged without the intervention and involvement of all the stakeholders concerned, in particular the local communities. Thanks to the Gelose law and the GCF decree, the State can transfer the management of these renewable natural resources to local grassroots communities.

2.2.1 Definition of natural resource conservation

Conservation is a concept recently used in the environmental field.

In its broadest sense, conservation refers to the act of preserving an element in a constant state. Nature conservation[74] consists of protecting populations of animal and plant species, as well as preserving the ecological integrity of their natural or substitute habitats (such as hedges, quarries, spoil heaps, ponds or other man-made habitats). Its aim is to maintain ecosystems in a good state of conservation, and to prevent or correct any deterioration they may suffer.

The English word "conservation" is translated into French as "préservation". In French, preservation explicitly excludes human beings (i.e. local populations) as the main cause of environmental degradation. In French, on the other hand, "conservation" is based on sustainable management in a variety of ways, and excludes neither local populations nor the measured use of resources. In French, there is no antagonism between conservation and development, i.e. commercial exploitation of resources within the limits of sustainable management.

Moreover, the concept of ecosystem does not exclude humans from its definition, since Tansley, who introduced the concept of ecosystem in 1935, already considered that "*human activity has its own place in ecology*" [Tansley, 1935, p. 303].

[74] In this context, nature *always* refers to biodiversity at the genetic, species and ecosystem levels and also, often, to geodiversity, patterning and other more general natural values. Geodiversity is the variety of rocks, minerals, fossils, topography, sediments and soils, and the natural processes that form and alter them. Geodiversity also means the variety of minerals, rocks (solid and loose), fossils, landforms, sediments and soils, and the natural processes that make up the earth's topography, landscape and underlying structure.

Furthermore, ecosystems and natural habitats cannot be conserved without managing a clearly defined, recognised and consecrated geographical area, by any effective means, legal or otherwise. This is a protected area[75] .

Protected areas are essential for the conservation of biodiversity. They form the cornerstones of virtually all national and international conservation strategies, and are set aside to maintain the proper functioning of natural ecosystems, to act as refuges for species and to preserve ecological processes that cannot survive in more intensively managed landscapes or seascapes.

Protected areas are the markers that enable us to understand the interactions between humans and the natural world. Today, they are often the only hope we have of preventing many threatened or endemic species from disappearing forever. They complement measures to achieve the conservation and sustainable use of biological diversity outside protected areas.

In the context of this definition, conservation refers to the maintenance *in situ* of ecosystems and natural and semi-natural habitats and of viable populations of species in their natural environments and, in the case of domesticated or cultivated species, in the environment where they have developed their distinctive properties.

The term environment refers to a forest. According to the FAO, a forest is land with tree crown cover (or equivalent stocking level) of more than 10 per cent and an area of more than 0.5 ha. Trees must be able to reach more than 5 m in height at maturity *in situ.*

A forest may consist of a dense forest formation where trees of different heights and undergrowth cover a significant portion of the ground, as well as open forest formations with a continuous vegetation layer where tree cover is greater than 10 per cent. New natural stands and all operational plantations that have not yet reached a crown density of 10 per cent or a height of 5 m are included in the term forest, as are areas that are normally part of the forest area but which are temporarily deforested as a result of human intervention or natural causes but which must become wooded again.

Forest includes nurseries or seed orchards that are an integral part of a forest; forest roads, clearings, firebreaks and other small clearings; forests in national parks, nature reserves and other protected areas, such as those of specific scientific, historical, cultural or spiritual interest; windbreaks and refugia belts with trees of more than 0.5 ha and a width of more than 20 m; plantations used primarily for forestry purposes, including rubber plantations and cork oak groves.

Forestry researchers are interested in "natural resources" or "forest resources". This concept is not neutral in the way the forest is represented. The French Academy dictionary defines the word resource generically as "*that which can provide what is needed*". The notion of resource therefore implies utility. In the case of natural resources, Weber et al [1990] point out that a species or thing in nature becomes a resource as soon as it is exploited by humans. The resource can therefore be seen as a unit located at the interface between humans and forest ecosystems. However, the *natural* nature of the resource encourages ambiguous use of the term.

---

[75] A protected area is a clearly defined geographical space, recognised, dedicated and managed, by any effective means, legal or otherwise, to ensure the long-term conservation of nature and its associated ecosystem services and cultural values.

2.2.2 Definition of the transfer of natural resource management

Management transfer means handing over the management of renewable human resources to local communities. A local community is constituted as a grassroots community (CoBa). This is any voluntary grouping of individuals united by the same interests and obeying the same rules of life. Depending on the case, it groups together the inhabitants of a hamlet, a village or a group of villages. The grassroots community has legal personality and operates like an NGO under the regulations in force.

Law no. 96-025 of 30 September 1996 on the local management of renewable natural resources, known as the "Gelose Law", sets out in a number of articles the resources to be transferred and the legal entity that will manage them.

The first article stipulates that, with a view to enabling the effective participation of rural populations in the sustainable conservation of renewable natural resources, the grassroots community may be entrusted, under the conditions set out in this law, with the management of some of these resources within its territory.

Article two (2) of this law explains that the renewable natural resources whose management can be entrusted to the grassroots community, under the terms of article one of this law, are those that fall within the domain of the State or the local authorities. of this law, are those that fall within the domain of the State or local authorities. This category includes forests, aquatic and terrestrial wildlife, water and rangelands.

The transfer is granted for an initial period of three years, renewable for further periods of ten years, after evaluation of the results.

2.2.3 Biodiversity conservation objectives

Madagascar's reputation for rich biodiversity is matched only by its level of poverty and the threat to its environment. In a document published by the World Bank entitled The Sad Analysis, GDP per capita fell from US$383 in 1960 to US$220 in 1990 and US$200 in 2000, placing Madagascar among the 15 poorest countries in the world. This increase in poverty has gone hand in hand with accelerating environmental degradation over the same period. The rapid disappearance of forest cover from 25% of the surface area in 1950 to 16% in 1995 is a case in point [ONE, 1995]. This situation highlights the close relationship between the environment and poverty. As a result, people are seen as the enemy of nature. With a very high rate of endemism, biodiversity must be conserved in order to :

- maintain genetic resources in domesticated organisms: preservation of these resources: immediate or potential use by humans (domesticated plants and animals + related wild species);
- conserving and perpetuating threatened animal and plant species: ensuring the survival of an endangered species and/or its long-term adaptability (evolutionary process);
- to preserve threatened ecosystems: the current nature reserves do not correspond in quantitative terms to the most threatened ecosystems;
- maintain the continuity of complex self-regulating communities, i.e. maintain their role in the functioning of ecosystems;

- make it possible to study and describe the diversity of the living world: understand (and monitor) the consequences of human activity on the functioning of species, communities and ecosystems.

Biodiversity is therefore threatened by human action. The destruction of ecosystems leads to the direct extinction of species. Natural resources that are only of economic value (food, medicines, fibres, fuel, etc.) are also in danger.

The proper functioning of ecosystems is essential to our survival (ecosystem services: oxygen production ; climate control by forests; recycling of bioelements, etc.).

2.2.4 The basis for conserving natural resources.

2.2.4.1 Decentralisation [RAMBINIZANDRY, 2004]

After decades of Jacobin centralism, the IIIè Malagasy Republic, under the impetus of France, particularly after the Baule speech, turned to a new form of administration: decentralisation. As a result, elected local authorities had to be given decision-making and management powers. These local authorities must adopt free administration.

Thus, in Title IV on the autonomous provinces, Chapter I, Article 126, the Constitution stipulates that :

"The Autonomous Provinces are public bodies with legal personality and administrative and financial autonomy...

The Autonomous Provinces, organised as decentralised territorial authorities, comprise regions and communes, each of which has a deliberative body and an executive body.

Similarly, the basis for decentralisation in the constitution of the current Fourth Republic is inspired by that of the former constitution mentioned above.

Article 3 of the IVè Republic stipulates that *"The Republic of Madagascar is a State based on a system of decentralised local authorities made up of Communes, Regions and Provinces, whose powers and principles of administrative and financial autonomy are guaranteed by the Constitution and defined by law"*.

As far as local authorities are concerned, article 139 of the current constitution stipulates that *"Decentralised local authorities, endowed with legal personality and administrative and financial autonomy, constitute the institutional framework for the effective participation of citizens in the management of public affairs and guarantee the expression of their diversities and specificities..."*.

The name and boundaries of each decentralised territorial authority may be changed by decree of the Council of Ministers, after consultation with the bodies of the Provinces concerned, on the basis of geographical, economic and socio-cultural viability criteria. The commune is therefore the basic decentralised territorial authority.

Once it has been appointed, the commune must get its administrative machinery up and running, otherwise it will be dismissed, so it will have the status of a *Fokontany*. To preserve its name, the commune must have a communal development plan, which will make it easier to negotiate with donors.

Also, in its subtitle II relating to the structures of the decentralized territorial authorities, article 149, the constitution of the IV^{è} Republic stipulates that:

"The municipalities contribute to the economic, social, cultural and environmental development of their area. Their competencies take into account constitutional and legal principles as well as the principle of proximity, promotion and defence of the interests of the inhabitants.

This means that it is decentralisation that gives autonomy to the Provinces, and even to the Communes. The lifespan of a Commune depends on its development, which in turn depends on the success of the powers thus assigned to the communes.

The *Fokonolona*, organised into *Fokontany* within the communes, is the basis for development and socio-cultural and environmental cohesion.

As a tool for managing and conserving renewable natural resources, the loi Gélose is based, par excellence, on decentralisation. By not adopting this decentralisation or empowerment, the transfer of the management of renewable natural resources did not exist either. The will to facilitate the transfer of renewable natural resources must be included.

Each decentralised local authority with natural resources must therefore facilitate their transfer to local grassroots communities wishing to manage them in order to ensure their sustainability.

Chapter Three: Theory of participation

3.1 Concept of Participation

The word participation seems to us to be a very simple word to which we don't pay much attention. According to Larousse, it refers to the fact of participating; but it also refers to the fact of associating with or taking part in an activity of some kind. Alternatively, participation can be the act of contributing to something in order to receive one's share. Also, to participate means to take part in decisions and/or to control the application of these decisions.

For more than a decade or so, specialists have been using this word to evoke a greater value in the accomplishment of an action. It has therefore become a method for involving participants in actions or fulfilling a share of responsibility.

This method is called the participatory method. It is based on the principle that interventions should always be planned and implemented with the participation of both women and men, and evaluated according to the extent to which they improve the living conditions of a population according to its own priorities.

This method has become very fashionable and has been adopted by international organisations working here, such as CARE and ADRA. WWF is currently adopting this method after noting a failure on the part of beneficiaries to take charge of projects.

In fact, the principle of this participation is to involve the grassroots community in the various stages of the project's implementation, from the study of its environment, to the actual design and implementation of the project, right through to its management.

Participation therefore takes several forms.

3.1.1 The different forms of participation [DEMONQUE M. & EICHENBERG J. Y., 1968]

3.1.1.1 Direct participation

Direct participation, as the name suggests, is where each individual participates personally. It therefore requires an area where the individual feels comfortable and integrated. However, participation can be subdivided into several types, such as :

✓ top-down participation.

This top-down participation is achieved by spreading power and responsibility downwards, because the top is not in a position to carry out all the tasks itself. This type of participation is most often found in the armed forces. This is why it is sometimes referred to as command participation;

✓ rising participation.

This participation also relates to participation in tasks, but it is different from so-called top-down participation. Top-down participation relates to the functions of the base, but does not prevent collaboration between the base and the top through the consideration of information emanating from the top;

✓ horizontal participation.

It consists of exchanging information concerning different sectors, acting in a common interest.

Horizontal participation is mainly based on information. This form of participation makes it possible to ensure consistency between these different sectors and to achieve objectives without internal dysfunction;

✓ general participation.

This participation concerns above all the circulation of information; and this circulation is not lacking. To this end, participants take responsibility for the information they receive on a regular basis regarding their activities. As a result, people receive the type of information they need according to their function.

3.1.1.2 Indirect participation :

It is also called collective participation. It is used when it is not possible to involve everyone in all phases of the action because the stakes are too high or there are too many different points of view, even if everyone concerned has been declared capable of participating on all points. As a result, the contribution is made with the help of representatives, who decide for the whole community. In this case, the relationship between the common interest and the particular interest is still clear.

This form of participation facilitates consultation and makes decision-making feasible.

3.1.1.3 Fruit participation :

It is a kind of profit-sharing for the work carried out. It is participation in productivity, profit and capital growth through self-financing. This participation in no way excludes the exercise of authority, the hierarchy of functions or the distribution of roles, but it does presuppose an uninterrupted dialogue between the top and the bottom. The forms of participation of each depend, in fact, on the structure and the objective of the company. To flesh out this explanation, we have also borrowed four other types of participation.

3.1.2 The different types of participation

3.1.2.1 Organised participation.

As its name suggests, this type of participation takes place within an organised framework and for explicit purposes. This type of participation can be conflictual, with two or more groups of participants having different ideas about the same problem. However, it does not imply an identity of opinions among all the participants.

3.1.2.2 Spontaneous participation.

Unlike organised participation, spontaneous participation takes place on a non-formal level, in small, spontaneously organised groups based on affinity.

3.1.2.3 Mandatory participation.

Participation is compulsory in the sense that certain standards have been set and must be met. For example, compliance with the terms of the contract between the agricultural contractor and his landowner. It is therefore more a question of compliance with rules. These rules are more or less coercive in nature.

3.1.2.4 Induced participation.

Participation is said to be provoked when a minority makes an effort to get the other members of the group to participate or to get non-members to participate.

3.1.2.5 Other types and forms of participation

Various researchers have defined the levels of stakeholder participation in any given process, but we should note the similarities between the categorisations of types of participation. The other forms of participation are summarised in the tables below according to the perception of the two researchers, Albert MEISTER and Silvain FORTIN.

Table 5: Participation according to Albert MEISTER

Type of shareholding	Origin and creation	Interests, activities, functions
De facto participation	Origins in traditional life : family group, religious group, profession...	Heritage conservation
Spontaneous participation	Created by the participants, it remains fluid, with no formal organisation: neighbourhood groups, amateurs, etc.	Meeting vital needs
Mandatory participation	Creation by external facilitators, mobilisation of the workforce for collective work	Grouping essential for the operation of the programme
Induced participation	Group provoked and generated as part of a project or programme: cooperative...	Adoption of collective behaviour and norms promoted by the intervening institution
Voluntary participation	Creation of the group as a result of an awareness, an initiative of the group itself	Satisfying needs, defending common interests, social advancement

Source: Albert MEISTER, 1974

Table 6: Participation according to Silvain FORTIN

Type of shareholding	Features
1. Activist participation	An individual's behaviour places him in a state of solidarity with other individuals
2. Integrative participation	It is an adjustment of a group's behaviour to pre-established norms.
3. Mobilising participation	Participants follow the instructions issued by one or more leaders
4. Consultative participation	Members are consulted on the development project
5. Participation in decision-making	Participants decide on their own development

Source: Silvain FORTIN, 1969

In natural resource conservation and/or management projects, voluntary and decision-making participation is what is sought because it is the culmination of an awareness-raising process and demonstrates the capacity of local stakeholders to initiate, lead and evaluate development actions in a concerted manner.

It should be pointed out that HIMO/VCT (High Labour Intensity)/(Live for Work) work is also considered a form of participation, because in the majority of cases, the population is remunerated with material rewards in response to their occasional help, known as "rewarded participation".

3.1.3 Customary framework for participation

If we refer to the classic definition of participation, which is the act of taking part in something, the analysis of the various forms of local participation according to the voluntary or compulsory nature of this participation makes it possible to categorise the type of participation as follows: the notion of participation is thus characterised in relation to the relations between two parties, namely the local players (often the local population) and the players 'from outside' the locality, who are often development projects or natural resource conservation programmes.

This view is shared by other studies, such as the one which argues that "we can then ask ourselves whether participation is not, primarily, an interaction between the role played by the agents of a development programme or project and the responsibilities imposed on all the members of the community involved in the development action".

The diversity of habits and customs across the regions of Madagascar: it is problematic to generalise about customary rights, which are likely to change from one locality to another. However, their similarity lies in the fact that each community has traditional regulations governing their locality. In terms of participation, for example, the literature refers to it as community solidarity or *"fihavanana"*. In fact, the *dina* or internal regulations/agreements of a community or *Fokontany* may require "participation" by all members of the community to show solidarity in work of common interest, failing which sanctions may be applied to recalcitrant members. It should be noted that Law 2001-004 governs the *dina* and sets out general regulations for the *dina* in terms of public safety.

3.1.3.1 Definition and objective

A participatory approach is an approach that will support development policies by involving the local population in the design, decision-making, implementation, monitoring and management of development actions. In other words, it gives the local population responsibility for its own development. It can be applied to all rural development programmes. Its main objective is to involve the local population closely in the design and management of all development activities in their area. Participation should not become an end in itself, but a means of building more appropriate and effective local development projects, such as :

- improve governance and encourage new or renewed forms of local democracy ;
- mobilise and involve local stakeholders more effectively through public debate and joint working ;
- gain a better understanding of local realities by drawing on the 'knowledge' that local players have of their area, so as to be able to think more effectively about the policy to be implemented;
- improve synergy between players and generate innovative and promising dynamics for the development of an area.

However, it should be borne in mind that the participatory approach sometimes takes forms that lead people to say that participation is "imposed and modelled" and that, at times, it leads to negative rather than constructive effects, but in any case, it does contribute to local development. Participatory development", in turn, is considered to be "a development process in which the initiative lies with the conscious and organised populations to carry out investigations and actions to promote their self-development".

3.1.3.2 Principle of the participative approach

The participatory approach is not an end in itself, but a methodological package. As such, it contributes to the socio-economic development of the area. The participatory approach actually tends to change the perception of the role of each of the stakeholders (State, local leaders, local populations, etc.), and to propose a sharing of responsibility between the different partners. The participatory approach is also a learning method based on experiential learning principles applied to adults, which promotes the sharing of personal knowledge and experience. It enables learners to participate effectively in constructing their own knowledge.

The participatory approach, a key tool for involving local people actively and responsibly, was born out of the failure of intervention strategies advocated in the past, as well as the government's fairly recent desire to incorporate the "people's participation" dimension into rural development policies. It supports efforts to decentralise technical services, disengage the state and privatise production and management activities.

It encourages local people to make decisions and take responsibility for actions designed to improve living conditions in their area. In other words, the aim is to associate and closely involve local people at the various levels and stages of the process.

3.2 Village participation

3.2.1 *Fokonolona* and Participation

Fokonolona comes from the word "*Foko*" meaning assembly or group and "*olona*" meaning human being. Technically, *fokonolona* means grassroots community.

It is a traditional institution that played an important role in the old administrative organisation, at least in the Merina country. It has an unwritten charter concerning its autonomy. Successive regimes in Madagascar have maintained it in different forms.

Under the transitional regime from 1972 to 1975, the Minister of the Interior set up the "*fokonolona*" for popular control of development. This *fokonolona* was identified by the implementation, through the "*dina*" established in a general assembly, of new powers and responsibilities in terms of administration, management of its assets, mutual assistance, hygiene and public health, conciliation and arbitration in civil matters and development.

Nowadays, especially during the crisis of 2002 and 2009, the *fokonolona* is called upon to play an important role in public safety through the *dina*. This role has been strengthened by the creation of the *Andrimasom-pokonolona (vigilance committee)*. Moreover, *the fokonolona* is a local administrative unit for social, economic and educational activities. This unit is chaired by the *filoham-pokonolona*. The structures created by the *fokonolona* will therefore be redeployed at all levels: good governance, family, economic, social, etc. This will contribute to the creation of a local environment that is open, healthy, stable and conducive to the optimal and efficient allocation of public and private resources, a sine qua non for social justice and dignity. The *fokonolona* will then become a prime area for rapid, harmonious and sustainable development.

The current world situation is changing because globalisation has become an inescapable system. To deal with this, the current government is trying to revive a modern-day "*fokonolona* institution" through legislation. These texts set out the general provisions on decentralised authorities: the "*fokonolona*" has become a basic administrative and economic unit, corresponding to the notion of hamlet or village in rural areas, and neighbourhood in urban areas. In addition, this entity was to play an important role in the social sphere.

However, the social relationships of the time of *Andrianampoinimerina*, which were based on "*fihavanana*", can no longer be applied in today's context, where people are becoming increasingly individualistic. The world is changing every minute, and yesterday's reality is no longer the same today. As a result, *fihavanana* has undergone a number of changes. Collective work is rare, and mutual aid and assistance are generally only practised within the family. The "*dina*" are ignored or applied less and less, for example: disinterest in community meetings is becoming noticeable, and the repression of thieves, particularly "ox thieves" in certain regions, is easing off.

This situation could be explained by the fact that the mutual aid of the "*valin-tànana*" era and community work have been replaced by the "employer-employee" relationship. The development of commercial relationships has destroyed the community. As a result, relationships have become superficial, governed only by individual interests. Major community projects can no longer be carried out in the same way as the idealised participation of *Andrianampoinimerina*'s time. What's more, a negative attitude on the part of the community towards public power is leading to a more serious conflict situation: it's difficult to find common ground, especially when it comes to maintaining public infrastructure or mobilising for public works, because people are becoming increasingly individualistic.

The democratic system was then introduced in Madagascar, with the aim of achieving development based on the Western model.

3.2.2 Democracy and participation

According to GRAWITZ, *"Democracy means the process of fostering equality by enabling the less privileged to have more rapid access to certain values: education, teaching, culture, responsibilities and the right to express their opinions*[76] *"*.

It presupposes majority rule, the freedom of individuals (respect for human rights) and the equality of citizens (which extends to equality of social conditions in socialist thought). But Abraham Lincoln, the American President, defined democracy as *"government of the people, by the people and for the people"*. This means that all citizens, regardless of birth, wealth or ability, direct or control political power. And there are different types of democracy, so we can distinguish between a political democracy that respects civic and political freedoms (freedom of expression, of the press, etc.) and an economic and social democracy that guarantees social rights (the right to work, to housing, etc.).

But what role does today's democracy play in citizen participation?

Democracy is seen as the most favourable structure for citizen participation, because in this sense, this form of citizen organisation reflects a power struggle.

Since the 1970s, a number of countries have been making the transition to democracy. The objective is clear: to seek respect for human rights and freedom in order to achieve the development they hope for. Democratisation takes many forms, from the violent overthrow of the government to reform from above.

Furthermore, according to Jean Jacques ROUSSEAU, liberal democracy should give the image of a citizen who must be interested in the affairs of the city and participate in political affairs, and who must have a strong motivation to participate in political life.

However, in today's society, especially during the crisis of 2009, participation in political life is very low and the population lacks motivation, especially during elections; on top of this, the lack of information on political issues only exacerbates the situation. Thus, during the crisis of 2002, the *Fokonolona* was called upon to play a role that was reinforced by the creation of a vigilance committee *called "Andrimasom-pokonolona"*. In a democracy, citizen participation is focused primarily on political participation. This participation generates control over life and the environment.

In short, this democracy and participation can be interpreted as one and the same thing, because the word participation is open simultaneously to the two meanings of the word democracy: the ideal of respect for people and all the institutional means that make it possible to approach it. Democracy therefore favours citizen participation from the outset.

3.2.3 Participatory democracy and GELOSE

[76] In Méthode des sciences sociales, Dalloz, 1996.

First of all, participatory democracy is the result of the failure of the Structural Adjustment Programme (SAP[77]) experimented with by donors in 1985. This programme was a policy devised and imposed by donors for indebted countries with a balance of payments deficit, especially those in the Third World. The aim of the programme was to finance indebted countries so that they could pay off their foreign debts and at the same time achieve long-term economic development. So the donors decided to draw up the programmes without the participation of the countries concerned.

In these activities, the population was seen as a mere beneficiary and not as a player in development. This situation led to a degree of inconsistency between the demands of the people and the actions taken, since the latter were focused primarily on macroeconomic reforms rather than the adjustment goals in question.

This incoherence, in turn, led to failure at state level, as it produced more or less disappointing results. These modest results were the fruit of the absence of any form of participation by the beneficiaries in both the design and implementation of the actions recommended by the Adjustment. The issue here is therefore one of incompatibility between the technocratic approach adopted by the donors and the needs of the people.

Moreover, the term "democracy", etymologically derived from the Greek *demos* meaning people and *kratein* meaning to govern. Democracy can therefore be translated, literally, by the following expressions: *government of the people or government of the majority.*

Democracy is therefore a system in which political power belongs to all citizens. The will of the people is considered to be the basis of power. In order to be participatory, democracy requires the foundations of participation. To participate is to take part or to experience oneself as part of a whole and to assume the active role that results from this belonging. Participatory democracy also refers to the notion of belonging and citizenship. Belonging justifies citizenship, which enables participation.

The concept of participatory democracy developed in France towards the end of the 1960s. Politicians' awareness of the limits of *representative democracy* led to a new concept of democracy. It gives greater consideration to public opinion. Scientific discoveries and technological advances were creating controversy in society. Citizens were given the means to express their opinions and influence the decisions that affected them. They were encouraged to debate, express their opinions and influence the decisions that concerned them. Participatory democracy is characterised by its deliberative nature. It encourages debate and discussion on matters of public policy by creating a platform for discussion between the players concerned. It goes beyond a simple democracy of opinion by focusing on issues of public interest.

In addition to the capacity for "deliberation", citizens also have power over their own development. The pragmatist philosopher John Dewey describes this role as a form of *"active and informed citizenship"* based on *"the formation of an active public, capable of deploying a capacity for enquiry and of seeking for itself a suitable solution to its problems".* This means

[77] The main objectives are: positive real per capita growth in order to raise the population's standard of living and reduce poverty; price and exchange rate stability; a more significant increase in tax revenues and control of public spending in order to reduce the financial operations deficit; regulation of currency in circulation according to the real needs of the economy; restitution of external foreign exchange reserves; compliance with external debt repayment schedules; gradual clearance of domestic and external payment arrears.

that citizens need to have access to a large flow of information so that they can understand their role and their duties in their development.

Decision-making power is always vested in the leader, with the aim of respecting the hierarchies and social order of a community. According to democratic values, this leader will be an elected official. Participatory democracy ensures that decisions take into account the needs of as many people as possible. Each individual is led to experience himself as a citizen. Being a citizen has two aspects: the exercise of rights and duties. In addition, they enjoy rights of expression, movement, development, and so on.

Madagascar's donors and leaders believe that to encourage the participation of grassroots communities in the conservation and/or management of natural resources, a policy of decentralisation and management of renewable natural resources is needed. This is known as "Gelose" or secure local management: local community management of renewable resources.

Gelose is a legal instrument established by Law 96-025 of 30 September 1996. It covers all renewable natural resources: forestry, inland fisheries, marine and pastoral resources. It sets out the institutions to be involved and the rules and conditions for developing and implementing the transfer of natural resource management. This law will gradually be supplemented by implementing regulations.

It is a mother law for the transfer of the management of renewable natural resources. It applies across the board to all resources whose management may be transferred. It is consistent with development policies and programmes and the legal framework for the transfer of renewable natural resource management. It sets out the procedure for implementing management transfer for the resources concerned: the players involved, the approach, the conclusion and the implementation.

3.2.3.1 Marginalisation and participation *Mahafale*

Liberalism reigns after defeating communism. As a result, it is becoming the dominant global system. This global system is gradually building the global village at a time when it is being torn apart by the global economic war.

In Madagascar, the national system acquired its sovereignty with independence, but it is internally torn between top-down and bottom-up dynamics (contradiction between *Fanjakana* and village). What's more, the tension between town and country is tearing the Malagasy dynamic apart.

Internally, particularly in the *Mahafale* plateau, the fundamental communities are neglected if not ignored. The communities have been split in two: rich and poor. The missionaries tried to wrest the *hazomanga* away from them, and the developers undermined the communities of descendants in favour of the acculturated elite, thus breaking up the family groups. Hence the marginalisation of village communities.

In concrete terms, marginalisation manifests itself on three fronts: the mystical front, the economic front and the human front.

Marginalisation on the mystical front is the fundamental marginalisation, i.e. the "*hazomanga* model"[78] has to be torn away, because for the developers, the first sign of a village's development is to tear away its *hazomanga*.

The neo-colonisers are trying to destroy the historical landmarks of the communities in order to replace them with another cultural identity. The missionaries tried to change the *hazomanga* into a bible, but this failed because the *Mahafale* are sedentary and their villages were not completely finished. The hamlets are scattered sporadically. What's more, most of the communities are illiterate. So cultural change seems a long way off on the *Mahafale* plateau.

On the economic front, the communities' last means of survival - the *raketa* (cactus) - has to be uprooted. They have replaced them with "*tsy mana baba"* biscuits. However, in a nod to the communities, some organisations such as the WFP have tried to organise the transplanting of cacti during the lean period using labour-based methods.

Then, on the third front, the members of the community itself are uprooted from their land. And so begins the "economic spoliation" itself; the compulsory departure, the entry into salaried servitude.

Thus, development as a process of violence leads to impoverishment on all 3 fronts. Nevertheless, it all starts with the uprooting of the founding way of living together, the *hazomanga* of the praying community.

At present, development means following the dominant model on a global scale. The poverty of the Malagasy people is no accident; it is the result of this development based on market logic, which is both modernising and, above all, oppressive. The tragedy of this development is that it kills all life in order to achieve development. The Malagasy people are demanding not just the first independence, but the true independence of a sovereign and no longer marginalised economy, yet colonial and neo-colonial violence has totally marginalised us from the nation, as each Malagasy person no longer makes history, no longer is the subject of historical initiative. We are on the sidelines, dispossessed, structurally excluded, destructured in every dimension of our existence.

The sacred and eminent dignity of Madagascan producers is being profaned. They are being commodified. This commodification makes it possible to monopolise producers and their products. This is what is known as economic alienation.

Mahafale communities are not immune to neo-colonial violence, despite their insubordination. Multinational companies bring superfluous goods and small radio stations to the weekly markets. The system put in place has encouraged the malaso phenomenon and this has weakened the agropastoralists.

In addition, international NGOs working in the field of biodiversity conservation have implemented the natural resource conservation programme by demarcating villages and seas and imposing a different management method from that traditionally applied by the

[78] For the *Hazomanga* model, the economy is essentially agropastoral. This means that :
A nomadic overland system, barely sedentary and therefore continually moving to temporary camps (*toby*) according to the rhythm of events or bovine transhumance, but villagisation is barely complete, followed by a system centred on bush farming. In short, the fundamental economy is that of beef, a bovine economy of transhumance also called contemplation or prestige, but this economy is total. In short, the fundamental economy is that of the ox, a bovine economy of transhumance also known as contemplation or prestige, but this economy is total.

communities. This indirectly pushes the communities of the *Mahafale* plateau to leave their ancestral lands. The inhabitants around the *Tsimanampesotse* park have had to give up their ancestral lands for conservation. They have had to move to other regions to pursue new activities. For those who have remained in the villages, whether on the forest resource management committee or not, they farm within the park.

As for the *Maroseragna* descendants of *Ampanihy,* they categorically refuse to co-manage the *Ankirikiriky* sacred forest *with* international environmental NGOs.

To conclude this chapter, participation is the act of taking part in any activity. In recent years, NGOs have used participation as a method of approach. This is known as the participatory approach. A closer look at the concept of participation shows that the principle of participation is to involve the grassroots community in the various stages of project implementation, from the study of its environment, to the very design, implementation and management of a project.

The Malagasy already have their own participatory approach based on community solidarity, *"valin-tanana"* or *"rima".* The approach differs according to the habits and customs of each region. However, donors are forcing us to apply another form of standardised participatory approach. Donors and the Malagasy authorities do not care about the diversity of customs and traditions across the regions of Madagascar. Standardising a participatory approach means generalising customary rights. This risks changing from one locality to another. This is why we can say that the participatory approach sometimes takes forms that make us say that participation is "imposed and modelled" and that it sometimes leads to negative rather than constructive effects, but in any case, for Madagascan decision-makers, it finally contributes to local development.

Conclusion part 1

To conclude this first part, we can say that the *Mahafale* landscape is located in the *Atsimo Andrefana* region. This region is known for its permanent aridity. At present, the *Mahafale* landscape is increasingly affected by the phenomenon of desertification linked to climate change. However, this *Mahafale* zone is home to a number of large forest blocks, the great reef and malnourished communities. Apart from this malnutrition, the *Mahafale* communities are destabilised by external domination. This domination is perpetuated by central government, evangelisation and schools, cultural domination, NGOs and donors. From within, the *Mahafale* communities are dominated by educated descendants who claim to be civilised. This has led to a widespread crisis, manifesting itself in various forms such as withdrawal, possession and rural exodus.

As far as the conservation of renewable resources is concerned, the Malagasy already had a policy for the protection of forest resources dating back to the feudal period. However, Madagascar did not have a marine conservation policy until after the signing of the Environment Charter in 1990. Strategies for conserving forest resources differed according to the period (feudal, colonial and current).

The communities of the *Mahafale* plateau already have their own forest management methods depending on the status of the forests (*"ala faly"* or taboo forests, *"alan-draza"* or ancestral forests). These forests are managed by the *ombiasy* and *mpitan-kazomanga* using the *"lilin-draza"* (laws of the ancestors).

Conservationists currently believe that the reefs and forest blocks of the *Mahafale* Plateau are exposed to various anthropogenic threats due to the practice of slash-and-burn cultivation of shrubby fallows, and they are imposing a standardised management method through the central government and international and national NGOs to conserve terrestrial and marine biodiversity.

In Madagascar, there was already a form of participatory approach based on *"valin-tànana"* or *"rima"* community solidarity. The approach differs according to the habits and customs of each region.

However, donors are forcing us to adopt a different form of participatory approach. Together with the Malagasy authorities, the backers do not care about the diversity of customs and traditions across the regions of Madagascar. They standardise a participatory approach and customary rights are generalised. This approach must change from one locality to another. Otherwise, participation is "imposed and modelled" and at times has negative rather than constructive effects.

However, we do not know whether these communities, which are extremely poor and dependent on these resources, will be able to upset the environmental balance and play an active part in conserving marine and land resources. This study will be the subject of the second part.

Part Two:

Conditions for participating in the conservation of forest resources

In this section, we will attempt to provide an analysis of the participation of all stakeholders and the conditions for their participation in the conservation of forest ecosystems.

Fourth Chapter: Participation in the conservation of forest ecosystems

For specialists, development is a set of social transformations that make autonomous, self-reproducing economic growth possible. At present, development follows a linear trajectory in which so-called developing societies are forced to "catch up" with developed countries, abandoning their identity in order to imitate them. This mode of development is destroying societies' capacity for innovation, their traditions, their social capital and, above all, their natural capital.

At present, the issue of economic and social development is combined with that of the natural environment.
natural environment issues, such as the depletion of the ozone layer, global warming and the loss of biological diversity,
global warming and the loss of biological diversity. This integration of environmental dimension in development issues has enabled human science to renew its original link with natural science in this specific field.

Sustainable human development is another form of development that emphasises this community identity. It is the development of man by man and for man, in which man is the principal actor.

In Madagascar, community development will start at the grassroots level by exploiting local natural resources.
natural resources. However, several ideologies have succeeded one another over time. time. The "hyper-conservationists" say that nature must be conserved, and to do so, it must be done without man, because man is considered to be the enemy of nature. For them, ecotourism must be developed to ensure the sustainable development of farmers.

For other ideologies: the systemic trend, for example, considers that man is part of biodiversity, so we cannot upset the environmental balance. The problem of man's relationship with nature is above all a special one. Nature cannot be conserved without a real effort on the part of society. With society, nature cannot be conserved without man.

As a result, the participation of these stakeholders is essential to harmonise and, above all, to achieve sustainable development.

4.1 Stakeholders involved in the conservation of natural resources.

The conservation of natural resources[79] and community development require the involvement of players at all levels. The first to be involved are those who finance the programmes and projects, then the intermediary players at national level, and finally the communities, which are the main players in the conservation of forest ecosystems to bring about a considerable change in their environment.

4.1.1 Financial backers or technical and **financial** partners

Donors operate at a global level. They persuade poor countries with their arguments on biodiversity conservation and sustainable development, using the collective participation approach. In this way, they encourage each country to increase its capacity to implement this process.

It has to be said that the term "donors" has now been changed to "technical and financial partners". They wanted to emphasise the technical side and hide the credit side from poor countries, whereas the ultimate aim of donors is to make a profit.

The main role of donors is therefore to contribute to the country's recovery through " civic" action, by advocating to governments the need for participation in development aid and cooperation programmes.

Madagascar's biodiversity is classified as a unique[80] and irreplaceable global public good. At global level, conservationists impose the rules of the game on poor countries through the international community. They persuade poor countries to apply for credit from the World Bank: this is known as environmental programme credit. Before accepting poor countries' requests for credit, the technical and financial partners impose their conditions and objectives.

In 1990, the promulgation of the Environment Charter heralded a change in the management of renewable natural resources: the disengagement of the State. In fact, the 90s were a period of socio-political upheaval in the country, culminating in the adoption of a general policy based on liberalisation of the economy, disengagement of the State from the productive sector and decentralisation.

Three facts have been noted with regard to the management of renewable natural resources. The first is that, due to a lack of resources of all kinds, the administrations in charge of this activity are increasingly powerless to stem the tide of destruction and illegal exploitation of the

---

[79] The *Mahafale* landscape is home to 34 species of reptiles and 57 species of birds, as well as numerous other endemic fauna resources: the lizards Ebenavia *maintimainty*, Oplurus fiherinensis and Paroedura *maingoka*, the geckos : Paragehyra petiti and Phelsuma breviceps, a species of snake: Madagascarophis oecellatus, birds Calicalicus rufocarpalis, Monticola imerinus, Pseudocossyphus imerinus, Anas bernieri and Voltzkowia *fotsifotsy*, a carnivore Galidictis grandidieri. The radiated tortoise Astrochelys radiata, the flagship species of the *Mahafale* landscape, where the highest population density is found in the Cape Sainte Marie reserve (3,000 individuals per km^2), and the birds Coua verreauxi and Coua Cursor populate the landscape in a representative manner. In the estuaries along the coastline, the mangrove acts as a veritable sediment filter and purifier of sea water, protecting the coral reefs and marine fauna.

[80] In EP III Project Document, World Bank: An impressive set of statistics testifies to the extraordinary richness of Madagascar's biodiversity: 99 per cent o f amphibians, 92 per cent of reptiles, 95 per cent of mammals, 83 per cent of plant species and 93 per cent of freshwater fish species are found nowhere else but in Madagascar. There are more than 1,000 known species of terrestrial vertebrates, 6,000 species of coral reefs, more than 12,000 identified terrestrial plant species and an unknown number of undescribed species. The country has been labelled the " eighth continent" in recognition of its unparalleled biological values.

nation's assets, because of the existence of open access situations. The second observation is that the populations living close to the resources are both victims and, to a large extent, responsible for the spiral of degradation of Madagascar's natural environment. The final observation is that village communities in several parts of the island apply certain forms of traditional resource management that are not valued because of the system of exclusivity applied by the administrations in this area.

Three workshops, including a national one, are being organised to decide what needs to be done within a legal framework.

In September 1994, the first national workshop was held in *Mantasoa* on the theme of "local management capacities and resource access rights". The workshop highlighted the need to give legal recognition to local natural resource management structures, despite the reluctance of the technical decision-makers present at the workshop.

The second workshop, which took place in *Mahajanga* in November 1994, was an international workshop. The theme dealt with was human occupation in protected areas. Ten points emerged from this workshop, signed by representatives of ANGAP (the current MNP), ONE and the Environment and Forestry Directorate. The most important of these decisions is the contractual approach making the State and rural communities partners in management.

The third was the international conference in *Antsirabe* in May 1995. This symposium was financed by the World Bank, USAID and the French Cooperation. The latter sent international environmental experts to guarantee the ONE document entitled *"Towards local community management of renewable resources: a proposal for a policy of decentralisation of renewable resource management"* in the light of trends in the international community concerning the conservation of biodiversity at world level. The aim of the conference is to validate the ONE document.

Fourteen points were declared. These take up the *Mantasoa* and *Mahajanga* concepts, in particular the inclusion of community management of natural resources within a legal framework. Almost a year after the conference, the Malagasy government promulgated law 96-025, known as the Gélose law, which was passed by the National Assembly.

Donors are also establishing partnerships with national and international NGOs to make programmes less political and, above all, closer to the grassroots.

This partnership is characterised by the financial support of technical and financial partners who provide financial support to make development programmes operational.

In order to publicise and implement their sustainable development projects and their prospects for conserving renewable natural resources, financial backers intervene within existing networks. Through these networks, they can actively intervene in problems by bringing together a number of people seeking to solve similar problems. In turn, these networks act as observers, but at the same time they convey ideas in order to propagate them. These networks may be governmental, corporate, academic or other.

When it comes to environmental protection in Madagascar, technical and financial partners do not give much thought to social and human development. By way of illustration, in its project document report relating to a proposal for an additional credit for the third phase of the environment programme in May 2011, out of USD 42 million, the World Bank earmarked just

USD 6.7 million for several activities: the creation of 900 local park monitoring committees, support for 30 park support committees around 30 protected areas, and for 150 community forest management groups (50 new groups to 100 existing groups) around 33 protected areas. To date, accompanying measures such as the construction of primary schools, the supply of drinking water, etc. are not on the agenda of Madagascar's financial partners in the *Mahafale* Plateau.

4.1.2 The State and its branches

4.1.2.1 Central administration

Given the current deteriorating situation in Madagascar, achieving participatory, sustainable development while conserving biodiversity is a real challenge for the entire nation.

Technical and financial partners are trying to help by setting up participatory development programmes through the State, which is seen as a gateway.

Consequently, the State must strengthen existing institutions for collective decision-making. The government is thus called upon to demonstrate its will and adopt an appropriate policy to enable collective decision-making.

The Malagasy government is obliged to commit to decentralisation and government reform under the banner of "Good Governance".

This new form of government management, based on a good relationship between the administration and the people it serves, gives rise to a new form of citizen access to decision-making. Good governance means participation. Madagascar is currently developing the means to promote this participation.

In the environmental field, decentralised natural resource management policies emerged in the 1990s. They were created in response to the failure of centralised policies to deal with forest degradation, and were encouraged by the growing international recognition of the ability of rural stakeholders to manage their forests. The application of these policies modifies the interactions between rural stakeholders and ecosystems in both positive and negative ways.

In fact, the Constitution and, above all, the texts relating to the implementation of renewable resource management transfers are clear on the stated desire for the effective participation of local people in biodiversity conservation.

This political decentralisation seems to facilitate participation at central level: it would give the local level the opportunity to take charge of the administration of renewable natural resources by developing a management tool for these resources and by actively involving grassroots communities in both decision-making and action. The grassroots community is therefore fully responsible for protecting the renewable natural resources in its area.

In Madagascar, land administration management is based on two systems for organising the administrative structure: decentralisation and deconcentration.

Decentralisation is a system for organising the administrative structures of the State which grants the transfer of financial, decision-making and administrative powers and responsibilities for undertaking service obligations from central government to local or regional authorities. In short, the principle of decentralisation is based on three transfers, namely the transfer of powers, the transfer of competences and the transfer of resources to decentralised local authorities.

Deconcentration is a system for organising structures in which the State transfers powers and responsibilities from central agencies operating in capital cities to the site offices of these agencies, within the State hierarchy.

However, the decentralisation policy in Madagascar raises a number of problems for the population, problems that could compromise national unity. And this has repercussions on the participation of citizens in the implementation of these programmes.

However, the State must fulfil its great responsibility. The State's responsibility is to ensure that the needs of the population are met. In general, to satisfy these needs, the State undertakes actions aimed at acting on these needs and directing them towards a deliberately privileged choice, given that needs are unlimited while means are scarce.

In the field of biodiversity conservation, this role mainly consists of encouraging individuals tobehave in a way that improves collective well-being, for example by applying laws and redistribution policies. A number of factors could lead to failures.

Firstly, the absence of an effective policy on the management of renewable natural resources can lead to failure. In fact, it is difficult to find a desirable, acceptable policy and to implement it, even though Law 96-025, known as the Gelose Law, followed by the GCF decree, has been promulgated. Trafficking in Madagascar's natural resources, such as precious woods and turtles, is on the increase.

The second failure of the State is its inability to match policy to the needs of the population. The demand emanating from the population is always changing, and differs according to social categories and their lifestyles, while State policy is inertial and bears no relation to this demand. This situation is combined with what Hamilton calls *"institutional drift"*. In fact, at the beginning, we can consider State policy as an institution serving to satisfy the population's demand, but it is only later that it becomes rigid because it is unable to satisfy this changing demand.

According to Hamilton, *"the life of an institution depends on its capacity to adapt...",* in other words, the effectiveness of the policy and the legitimacy of the state depend on its ability to satisfy any demands that may arise. However, the state is unable to develop a policy that satisfies the demands of the population, either in terms of income or well-being, but also the need to manage the environment.

The State is also faced with networks of traffickers in natural resources. The illegal export of precious woods such as the famous rosewood and precious stones is a case in point. However, it is difficult for the State to control and enforce the laws on traffickers because some leaders are in league with those who illegally export these resources, especially those listed in the categories (banned from export) of the CITES convention. We all know from the media (newspapers and radio) that the representative[81] of the State or those close to him are involved in the rosewood or precious stone business. The State is currently between a rock and a hard

[81] The Anjozorobe precious stone affair mentioned that a quarry is monopolised by Masoandro and his son. The term Masoandro or sun is military jargon for the supreme leader of the army, who is none other than the Head of State.

place. It is weak in the face of the Convention and the sanctions that CITES will apply to it, but it is at an impasse in the face of pressure from traffickers.

It is in the interests of each country to declare what the environment can do for it and what it is actually destroying. For example, in the case of greenhouse gas emissions, each country has an emissions quota that must be met or face penalties, but the country in question is not meeting its quota. When questioned, it underestimates its emissions, even though the impact on the environment has always increased. This misrepresentation is particularly common in industrialised countries. This phenomenon is exacerbated by the inadequacy of international institutions that can take decisions accepted by all, and the lack of a global control system.

The failure of the State to manage natural resources can also be due to the contradiction between environmental management policies and the agricultural development policy undertaken by the State. For example, on the one hand, the state has drawn up a forest protection policy, but on the other, it is encouraging farmers to adopt an agricultural-based economy.

In this case, to encourage the economy, the state can grant subsidies to farmers or guarantee the price of production to encourage producers to increase their output. To increase production, farmers have to over-exploit the soil, and if there is not enough arable land, they have to look for more. Hence the proliferation of slash-and-burn farming. This is known as "subsidised destruction". The State is therefore forced to make a choice between the well-being of the population and economic growth, and it should be noted that this situation is only valid if farmers are motivated by offers from the State.

4.1.2.2 The region

Article 141 of the Constitution of the Fourth Republic[è] stipulates that *"the decentralised local authorities, with the support of the State, are responsible in particular for public safety, civil defence, administration, regional planning, economic development, environmental protection and improving the quality of life.*

In these areas, the law shall determine the allocation of powers, taking into account national and local interests.

Article 2 of Law No. 2004-001 of 14 June 2004 states that*"the Regions are public authorities with an essentially economic and social vocation. They direct, stimulate, coordinate and harmonise the economic and social development of the whole of their territorial jurisdiction and, to this end, ensure the planning and development of the territory and the implementation of all development actions.*

As a result, the region has multiple areas of competence, such as :

- control of the legality of the acts of municipal authorities;
- identifying the region's priorities;
- drawing up a framework programme and/or a regional development plan;
- drawing up regional land use plans;
- preparing the programming of regional development initiatives;
- including hydro-agriculture, fisheries, small-scale industrial and commercial promotion, promotion of the services sector, livestock farming, etc.

All of which goes to show that the region has a major responsibility in environmental protection and local development. Despite the efforts made by the government, the 22 regions are finding it difficult to get started. The causes of this difficulty are: inconsistencies in the texts that govern them and the excessive reduction in their material resources.

What's more, the status of the regional leader is itself ambiguous. Currently appointed by the government, the head of the region will eventually be elected by the regional council. The latter has already been set up and we are currently awaiting the election timetable. The activity of the regional chief is currently being closely monitored by central government. He combines the powers of local representative of the State and head of the regional executive, with the result that there is confusion over the region's powers over devolved territorial services.

In addition, the administrative divisions of these departments do not correspond to the divisions of the decentralised local authorities, hence the confusion.

Their material resources are limited, even though these local services are obviously essential in the fight against poverty and to improve living conditions. There are also many contradictions between regional and local priorities, as the regional plan is not simply the sum of local plans.

Most regional chiefs are only aware of the existence of community management of renewable natural resources in the event of a dispute between communities. Environmental NGOs rarely report on their activities to the region.

4.1.2.3 The *Commune*

Article 149 of the Constitution of the IVè Republic stipulates that *"The communes contribute to the economic, social, cultural and environmental development of their territorial jurisdiction. Their competences essentially take into account constitutional and legal principles as well as the principle of proximity, promotion and defence of the interests of the inhabitants"*.

The Commune is therefore the main pivot of local development. It provides all local services, as well as development initiatives. These resources, both human and financial, are still weak, particularly for rural communes. State grants and subsidies are still mediocre and irregular; own resources are difficult to mobilise.

The municipality is fully responsible for managing all matters of municipal interest. This is reflected in its concern to improve the well-being of its residents and, where appropriate, of all those who use the municipal area.

The local authority has a very wide remit. Secondly, it acts as an intermediary between the State and citizens for certain administrative formalities.

The municipality carries out a number of autonomous actions:

- Directly manage its development and planning, using the municipal development plan. Issue the various permits (building and demolition permits, etc.);
- be able to join forces with various partners in major development projects;
- provide local public services (lighting, roads, drinking water, etc.);
- be able to play a complementary role in a wide range of areas such as social assistance, the local economy (business start-ups, etc.) and the promotion of social, sporting and cultural life.

Generally speaking, there is no transparency in the management of the commune. The low technical capacity of local elected representatives is an obvious source of difficulties.

When we look at the 13 communes that make up the *Mahafale* landscape, we see that decentralisation is taking place in theory, because administrative decentralisation is not accompanied by the transfer of resources. In fact, the transfer of new powers obliges the communal authorities to find the resources they need to operate internally. The concern of the communes in the *Mahafale* landscape is how to increase communal revenues.

To increase its revenues, the municipality has two choices: either it increases the rates of these revenues, or it adopts a policy that favours an increase in the tax base. Whichever choice the municipality makes, it is always the communities or taxpayers who will have to bear all the costs.

The leaders of the 13 communes leave the management of their forest resources to the community and to the biodiversity conservation support NGOs operating in their commune. The mayors simply sign the CoBa files to authenticate them. The commune's leaders do not even attend the demarcation of the land to be transferred to the communities. "*We don't know much about this management transfer, so we don't do much about the activities undertaken by the CoBa*" [Interview with Deputy Mayor C.R *Itampolo*].

4.1.2.4 The ***Fokontany***

As for the *Fokontany*, Article 152 of the Constitution of the Fourth Republic stipulates that *"The fokonolona, organised into Fokontany within the communes, is the basis for development and socio-cultural and environmental cohesion. The leaders of the Fokontany participate in drawing up the development programme for their commune"*.

The *Fokontany* Chief is therefore responsible for the general administration of the *Fokontany*. The other members of the committee (vice-chairman of the *Fokontany* and the mobile districts) assist in the exercise of these functions in accordance with the procedures laid down in the implementing regulations.

The *Fokontany* Committee participates and contributes in a permanent and effective manner, under the authority of the *Fokontany* Chief:

- *Fokontany* development activities;
- educational, sporting and cultural activities of the *fokonolona* ;
- social or community mobilisation and health activities;
- environmental protection activities and risk and disaster management;
- day-to-day management and protection of public infrastructure and works;
- electoral or referendum work and operations;
- communicating events of all kinds concerning the *Fokontany* and its environment;
- the application of *dina*.

In addition, it carries out the specific duties entrusted to it by the Mayor, executes and finalises the Mayor's instructions and directives and assists the administrative and judicial authorities in preventing and punishing offences, particularly acts likely to disturb public order.

The *Fokontany* chief is responsible for implementing the security plan in the *Fokontany* under his jurisdiction. They draw up and implement local security plans with the support and contribution of the *Andrimasom-pokonolona*.

In the case of the *Mahafale* Plateau, the *Fokontany* chief and committee are actively involved in the process of developing management tools for the transferred land and in the process of transferring the management of renewable natural resources, particularly with regard to the demarcation of the land and the creation of the *dina* that will govern the management of the land transferred to the grassroots community. Most of the *Fokontany* committees or chiefs join the CoBa as full members. Moreover, most of them are descendants of the *Mpitan-kazomanga* of their neighbourhoods.

In short, the involvement of the central and regional administration in the conservation of renewable natural resources is very low. However, local administrative structures such as the Commune and the *Fokontany* have important roles to play in the local development process as well as in the protection of biodiversity because these communities know the reality in situ. They are capable of involving citizens in the development and conservation of biodiversity in their locality, but the success of this depends of course on their ability to motivate the grassroots community.

According to the World Bank's environment report, Madagascar's natural resources are not optimally managed. This mismanagement is costing the country money, whereas one of the main challenges facing grassroots communities is economic and social development to ensure that this world heritage is properly managed.

In its report on the environment (Country Environmental Analysis) entitled: *Key Messages*, published in 2013, the World Bank commented on Malagasy public spending in the following terms: "*public spending in the environment sector (restricted to the Ministry and the two agencies*[82] *mentioned above) is around 10-20 million dollars per year, or 1‰ of GDP since the crisis, 2‰ before this crisis (or 1% of the national budget). This level is low compared with the standard of 1% of GDP, the cost of environmental degradation, and above all the fact that the country has one-twentieth of the world's biodiversity*".

The content of this report reveals the extent to which the State and its agencies are involved in biodiversity conservation or environmental protection in Madagascar. If the State and its agencies do not demonstrate their willingness to protect renewable natural resources, the efforts made by grassroots communities will be in vain.

4.1.3 NGOs as intermediaries for donors

"*Tao trano tsy efan'irery fa ny entan-jarai-mora zaka*", literally means that building a house is not a personal matter, but a burden becomes light if it is carried by many people. The philosophy of this adage of our ancestors is borrowed by donors to accomplish their mission.

They work at grassroots level, sometimes with the help of intermediaries. They therefore establish partnerships with national or international NGOs and associations. This approach makes the programmes less political and closer to the grassroots. This partnership is

[82] MNP and ONE

characterised by financial and technical support to make development programmes operational, but above all to justify its interference in public affairs.

o enter into a partnership contract with donors, associations or NGOs must draw up a project document and submit it to donors; those with the same objectives as the donors, and especially those with a great deal of experience, are accepted by donorsas partners in the implementation of the project.

To help achieve the Durban vision, the WWF-French Global Environment Facility (FGEF) co-funded the COGESFOR project through the French Development Agency (AFD). The project is entitled *"Sustainable management of natural resources for the conservation of Madagascar's three biodiversity hot spot regions"*.

The aim of the project is both to protect Madagascar's biodiversity and to reduce poverty, by setting up participative and sustainable natural resource management frameworks.

It is part of the continuity of pilot operations for local management and biodiversity enhancement carried out by the FGEF since 2000 in Madagascar. It is consistent with the national policy of transferring management, based on the principle that a certain amount of development can be a prerequisite for the conservation of fragile areas, and with the development of the Madagascar Protected Areas System (SAPM).

The project has three main components:

- concerted development, based in particular on the transfer of management to local communities (*VOI*), supported by secure land tenure in the areas concerned;
- the development of environmentally-friendly value-adding sectors (wood-based sectors, aromatic and medicinal plants or other promising sectors);
- setting up decentralised, self-financing forestry control systems.

The project pays particular attention to the biological and socio-economic effects and impacts it is likely to have: a monitoring system will be designed and set up, then transferred to the national section.

The organisational approach adopted is based on collaboration between different partners: CIRAD, WWF and various NGOs. The project has an operational unit in *Antananarivo*, bringing together the various players on an equal footing, with two local offices in the selected hot spot regions. Particular attention is being paid to building on the achievements of previous FGEF projects and ensuring that there is a genuine exchange of experience.

The project is focusing on three target regions, characterised both by their unique biological wealth and by the serious degradation of their natural resources as a result of human practices. These target regions are :

- the *Atsimo-Andrefana* region, with the *Tsimanampesotse* National Park and the *Mahafale* plateau;
- the *Menabe* region with the *Kirindy Mite* National Park;
- the *Alaotra-Mangoro* region, with the recent creation of the *Mantadi Zahamena* corridor, paving the way for the creation of new protected areas (NAPs).

The *Mahafale* Plateau is one of the priority areas for biodiversity conservation in the WWF's *Ala Maiky* eco-regional programme. It is under heavy pressure, particularly from agricultural clearing for maize production and the manufacture of charcoal used in towns as domestic fuel. Implementation of the COGESFOR project in the *Mahafale* plateau is therefore the responsibility of WWF.

Each NGO working in Madagascar has its own way of operating. It is therefore necessary to understand how WWF operates in Madagascar before assessing its involvement in the conservation of forest ecosystems.

How WWF Madagascar works[83]

WWF Madagascar is a Regional Representation Office. It manages a large number of projects in the Indian Ocean region. The procedures for awarding contracts, committing funds and financial reporting are extremely strict to guard against any criticism in this respect. Staff members are trained in this rigorous financial management (grant release, production of financial reports and accounting audits).

The main complication is in the day-to-day management of projects. For various reasons (in particular the capacity of the staff working in the branches), all calls for tender, selection of service providers, recruitment of staff and signing of contracts are managed by the WWF *Antananarivo* office. Similarly, all commitments/expenditure of less than MGA 3 million (around €1,200) must be approved by the ecoregional leader and the *Ala Maiky* programme manager, those of between MGA 3 and 10 million (around €4,000) by the Conservation Director, and those of more than MGA 10 million by the Regional Representative.

Contractual and financial monitoring therefore requires too much to-ing and fro-ing between the WWF offices in *Antananarivo* and *Toliara*, and this "two-headedness" in project coordination leads to a certain amount of confusion in the management of files (who manages which contract? who is really in charge of monitoring which study? what is the level of disbursement per action?).

Clearly, this cumbersome administration hampers the progress of activities, not only at the level of the project unit, which has very limited autonomy, but also and above all at the level of the service providers, for whom the procedures imposed by WWF are considered too cumbersome and difficult to apply in their context. By way of example, it should be noted that the project has had to take charge of financial management training for service providers (application of the WWF procedure manual). While this activity can indeed be considered as capacity-building for the local NGOs with which the project works, its effectiveness is doubtful insofar as the high *"turnover"* (change of management) observed among these service providers means that they are unable to capitalise on this training.

Although it helps to overcome the cash shortages of small service providers, the *"cash-flow"* system used by WWF also constitutes an additional constraint for operators insofar as successive financial instalments are only released on receipt by *Antananarivo* of all the accounting documents verified by the Administrative and Financial Office in *Toliara*. Since, despite the procedures, the methods of checking seem to differ between local and national

[83] In Conservation et Développement du plateau calcaire *Mahafale*, ex-post evaluation of the FFEM project/valorisation des acquis et opportunités, June 2007.

managers, this unfortunately leads to a lot of back and forth between the service provider (if there is one), WWF *Toliara* and WWF *Antananarivo*, which delays the proper execution of activities.

In extreme cases, we might even be led to believe that the failure to carry out certain activities (training local agents and operators, for example) could be attributable to these procedures, which are deemed too cumbersome by the local antenna team responsible for implementing them.

The procedures applied within WWF are delaying the implementation periods for the activities mentioned in the project. The implementation and, above all, the monitoring of project activities require sufficient human resources. To undertake the COGESFOR project, WWF dedicates only 2 people: a project manager and a socio-organiser. The latter was replaced after 7 months of work for bad behaviour towards the communities. The branch then recruited 6 local contact agents who were none other than members of the communities belonging to the AICPM. They have not received the necessary training in the protection of renewable natural resources.

Until the end of the project, 25 CoBa will be set up around the *Tsimanampesotse* park: 15 new management transfers and 10 renewals.

The implementation of CoBa management tools (specifications, management plans, etc.) is behind schedule. Given the lack of capacity building, the application of the dina is uncertain. The implementation of the CoBa ended ten months before the end of the project. This delay implies a lack of technical support and joint reflection with the beneficiary grassroots communities on the aspects of local tax incentives, flow control and land tenure security.

4.1.4 THE AICPM

AICPM or the Intercommunal Association for the Conservation of the *Mahafale* Plateau is an association created by AGERAS in 1997 with the aim of protecting *Tsimanampesotse*. Initially, it was a cooperative governed by law 99-004 and involved only 5 communes around the *Tsimanampesotse* park. In 2009, following an evaluation by WWF, MNP, SGP and *Tany Meva,* its status was changed. It became an association governed by law 60-133, but its purpose remained unchanged. The number of member communes has now reached 13.

Each commune is represented on the AICPM by 2 people, and the association's areas of activity are bounded to the south by the *Menarandra* river, to the north by the *Onilahy* river, to the west by the Mozambique canal and to the east by national road no. 10. The mayors of the 13 communes are considered to be ex-officio members of the association, and each commune contributes 20,000 *ariary* per year. It should be emphasised that the AICPM works with the AVG at national level, particularly in the fight against land turtle poaching.

The AICPM is divided into 5 sectors:

- the southern coastal sector comprising the municipalities of *Androka* and *Itampolo* ;
- the northern coastal sector through the communes of *Beheloka, Anakao* and *Salary* sud ;
- the RN10 north sector through the commune of *Betioky* ;
- the central RN10 sector through the municipalities of *Beahitse* and *Ejeda*, and

- the RN10 South sector through the communes of *Ankiliabo* and *Ampanihy*.

AICPM has a *dina* that encompasses those of associations working in the environmental field in its area of intervention. WWF, MNP, GIZ and SGP/Tany *Meva* fund the association's activities relating to the conservation of the *Tsimanampesotse* park, as well as grassroots community development activities in the *Mahafale* landscape.

Since the AICPM began implementing alternative activities for the conservation of natural resources, it has been unable to complete the activities it has undertaken because of misappropriation by the central office.

In general, the participation of environmental NGOs and associations in Madagascar does not support community development. Rather, it enriches the members of the associations or NGOs. However, the technical and financial partners have included in their objectives for phase three of the environmental programme a contribution to improving the quality of life of the population.

For some time now, Madagascar has been a playground for NGOs and associations. Most of the promoters of these NGOs or associations do not come from the countryside. The promoters know that the technical and financial partners serve the NGOs when the donors are aiming for the development of grassroots communities, because the latter are involved but not concerted.

The development pronounced by technical and financial partners therefore does not serve the populations most directly concerned, who should normally have a say in their own development. Development has therefore become a real business. As a result, there are a lot of people (urban elites) and strategies competing for access to a certain project. It's a real international competition between NGOs.

NGOs are ultimately the development players. They are often the prime contractors for much more global projects. They are therefore funded by international donors, the state, etc. They are the ones who will set up the projects at local level. They are the ones who implement the projects at local level. So it's the NGO workers who do the work, and they have to be as sensitive as possible to the realities they encounter in the field, but are they trained to do this? And do they have the means to do so? Do they have the time to do it, even though their work is often just as essential?

The answer is no, because NGOs and associations are financially dependent on donors. Without them, they can do nothing. This ignorance leads them to tailor their work to the expectations of the donors who commission them, and disconnects them from farmers' realities.

This situation of dependence encourages these NGOs or associations to be neither innovative nor critical in their approaches and working methods.

As a result, the staff of NGOs and associations working to conserve natural resources either adhere to global environmental standards or keep their disagreement to themselves in order to protect their own interests.

There is also a lack of discussion between all the NGOs involved in the environment or within the NGO itself. This shortcoming is also due to the inadequacy or non-existence of evaluation of the action of NGOs or associations in the field. It is true that they send in annual reports showing what has been achieved in relation to the initial objectives. Their financial audits testify to the seriousness of the management of funds, but they have rarely drawn up a qualitative assessment of the socio-economic and environmental impacts.

4.1.5 The grassroots community

The grassroots community is the basis for sustainable human development, so that it can take charge of its own development.

The role of local elected representatives (mayors, etc.), civil society (associations, etc.) and individual members of the grassroots community is to monitor the responsibilities of the State or decentralised local authorities in the management of public affairs, but above all to ensure the self-governance of the grassroots community in a participatory manner.

In general, members of the local community are the most familiar with the realities in which people live and the problems they face.

With regard to participation in the commune's development programmes or the conservation of natural resources, the grassroots communities organise themselves at *Fokontany* level to form an association to manage the renewable natural resources in their village.

The structure that works at village level is called the *Vondron'Olona Ifotony* (*VOI*) or grassroots community. It is governed by decree 2000-027.

25 CoBa have been set up around the *Tsimanampesotse* national park with a dual objective: to protect the forest ecosystems around the park and, above all, to conserve the biodiversity of the *Tsimanampesotse* park.

CoBa members are appointed by the CoGe and the General Meeting respectively.

- CoGe (Management Committee) or Bureau

The CoBa office structure is as follows: a chairman, one or two vice-chairmen (depending on the number of *Fokontany* affected by the natural resources to be managed and, above all, willing to join the CoBa), a secretary, a treasurer, councillors (there is at least one councillor per*Fokontany*), two auditors and a village development coordinator.

- General Meeting

This assembly is made up of members. Membership of CoBa is subject to the following conditions: the person concerned must be of political majority, i.e. 18 years of age or over, he or she must live in the village where the management transfer takes place, and he or she must pay the membership fee and monthly dues.

With the exception of the *Ankara* CoBa, namely the *Fokontany* of *Marofotsotse (Mitsinjo Taranake Magnasoa Tane)*, the *Fokontany* of *Behomby (Magnasoa Tane),* the *Fokontany* of *Ankitekiteke (Fiarovantsoa)* and the *Fokontany* of *Vorojà (Tsivery Anjara Mahasoa),* all the CoBa in the *Mahafale* landscape are made up of at least one *Fokontany*. The number of members per CoBa varies from 10 to 522.

Map 8: Location of CoBa in the *Mahafale* landscape

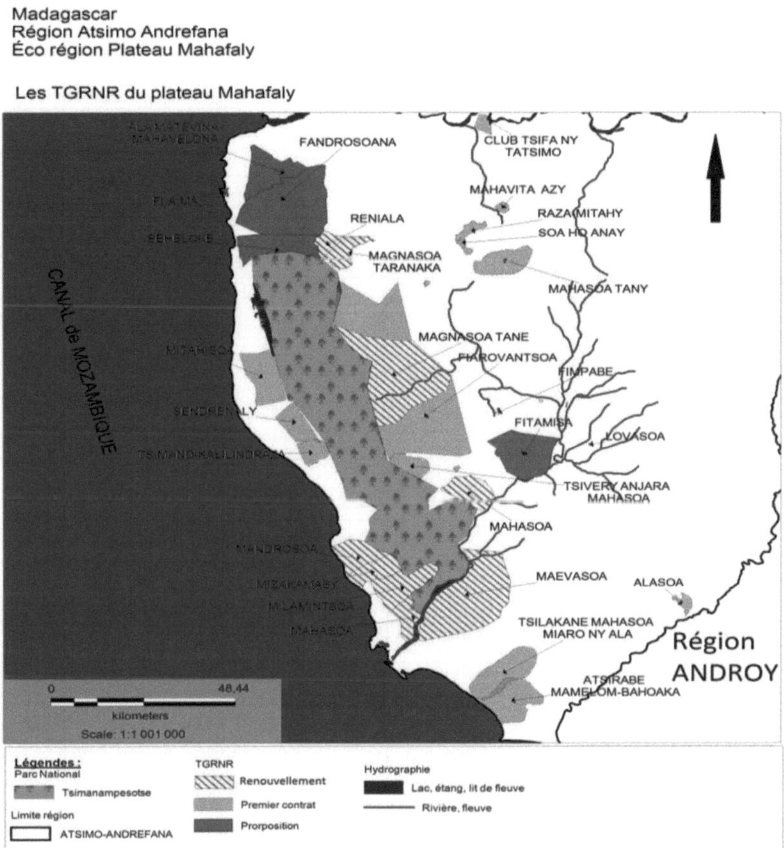

Source: COGESFOR *Toliara, 2014*

Table 7: Data by CoBa in 2013

N°	Name of CoBa and *Fokontany* headquarters	Creation date	Number of member *Fokontany*	TGRN surface area in Ha	Type TGRN	Number of members
01	*Mizakamasy (Nisoa)**	September 2005	4	6 980	GCF	522
02	*Mandrosoa (Tongaenoro)**	September 2005	3	8 271	GCF	300
03	*Maevasoa (Zamasy)**	December 2006	5	38 794	GCF	267
04	*Mahasoa (Ambalatsimiviky)**	September 2006	10	6 636	GCF	165
05	*Milamintsoa (Befolotse)**	December 2006	4	11 130	GCF	522
06	*Miovasoa (Beheloke)*	June 2009	3	9 600	GCF	201
07	*Fandrosoana (Ankilimivony)*	March 2011	4	17 600	GCF	59
08	*Sendrenaly (Vohombe)*	December 2010	2	7 440	GCF	31
09	*Mitahisoa (Antanindranto)*	December 2010	2	12 562	GCF	21
10	*Tsimandikalilindraza (Bemanateza)*	December 2010	2	4 568	GCF	26
11	*Ala matevina mahavelona (Andranotohoka)*	November 2012	3	12 996	GCF	85
12	*MIAHI (Miaro ny Ala sy ny Horake eto Itampolo) (Itampolo I)*	March 2005	1	500	GELOSE	17
13	*FIVELOMA (Itampolo II)*	March 2005	2	1 125	GELOSE	12
14	*Mahazoarivo (Malangiriake)*	March 2005	1	900	GELOSE	10
15	*FIMPABE (Bekinagna)**	December 2006	1	3 554	GCF	143
16	*Mitsinjo Taranake (Itomboina)**	October 2006	5	17 736	GCF	123
17	*Reniala (Ampotake)**	October 2006	1	11 856	GCF	38
18	*Miarimpototse (Ekelelahy)* (Ekelelahy)*	December 2006	1	18 062	GCF	150
19	*Magnasoa Tane (Behombe)**	September 2006	1	42 999	GCF	45
20	*Mahasoa (Ampitanake)**	December 2006	1	8 664	GCF	105
21	*Magnasoa Tane Mitsinjo Taranake (Marofototse)*	March 2011	1	23 550	GCF	68
22	*Fiarovantsoa (Ankitekiteke)*	March 2011	1	14 552.3	GCF	62
23	*Soa Ho Anay (Andremba)*	March 2011	7	1 052	GCF	159
24	*Tsivery Anjara Mahasoa (Vorojà)*	March 2011	1	3 987	GCF	220
25	*FITAMISA (Sakoantovo)*	December 2012	2	18 060	GCF	122
	Total		68	245623	3 GELOSE 22 GCF	3473

Source: Our own survey, 2014

* CoBa manager of former management transfer under the GELOSE contract having renewed their contract in GCF.

We note in this table that 3 CoBa are still under the GELOSE management transfer contract. They were supported by SAGE. Since the project to transfer the management of SAGE's renewable natural resources ended before the end of the CoBa management contract, the CoBa have not yet been assessed by the forestry administration. SAGE did not have the opportunity to entrust its CoBa to the support of other promoters. Moreover, the CoBa supported by SAGE manage caves, whereas those that manage forests are of interest to other promoters such as WWF.

As for the following CoBa: FIMPABE in *Bekinagna*, *Mitsinjo Taranake* in *Itomboina* and *Reniala* in *Ampotake*, are supported by GIZ by changing the contract to GCF.

The *Tsimandikalilindraza* CoBa in *Bemanateza*, Mitahisoa in *Antanindranto* and *Sendrenaly* in *Vohombe* are supported by MNP. Those not listed are supported by WWF.

The total surface area covered by the TGRNR varies between 1,052 ha and 42,999 ha. The surface area of each zone varies from one CoBa to another. The first 14 of the list of CoBa in this table border the coast. In total, 245,623 ha of forest have been demarcated for 3,473 members spread across 68 *Fokontany*. 1 to 10 *Fokontany* members to manage forest resources.

On the coast, the motivation for the *Fokontany* to join a CoBa is to be able to benefit from the use of the user rights zone, as the conservation zones of the first TGRNRs (CoBa marked with an asterisk) initiated in 2005-2006 are for the most part integrated into the park. The current TGRNRs retain virtually only the cultivation, grazing and right-of-use zones that merit increased regeneration due to the reduction in available space.

The extension of the *Tsimanampesotse* park has obviously led to a reduction in traditional access to forests, while the use rights of coastal communities have been exhausted and people have only wooded savannas to manage and/or use.

Photo 4: *Ambohitse* forest managed by the *Tsimandikalilindraza* CoBa *(Bemanateza)*

The first TGRNs are management transfers from CoBa, which entered into a contract with the State under the Gélose contract during the FGEF I project and completed their 3-year management contracts. After the evaluation, they received a favourable opinion from the forestry administration to renew the ten-year contract. FIMPABE, *Reniala* and *Mitsinjo Taranake* were renewed by GIZ and the rest by WWF.

4.1.5.1 The composition of the bureau

There has been an improvement in the status of the CoBa compared to the former.The article on
the term of office of the officers is the most affected by this improvement. The officers are elected by the members of the CoBa for arenewable term oftwo years, whereas the old statutes stipulated that the officers performed their duties until the end of the management contract, i.e. for 3 years. The members noted the various kinds of abuses committed by the management committees during the first contract and reduced the term of office of the executive committee.

Those newly managing their land have also followed the experience of previous transfers. They have limited the mandate of an office to two (2) years and if it is renewed, it will be responsible for renewing the second contract, which will last 10 years.

To uncover other hidden problems linked to CoBa's existing structure, we will try to find out more about the management style adopted by the committee and who makes up the offices?

Thus, studying the composition of the boards helps us to understand whether the transfer of management leads to an investment by outsiders in the village community or reinforces the existing power structure that the members have denounced as abusive. In both cases, the system is asymmetrical and tensions arise between these powers: foreigners versus indigenous groups, strong groups versus socially less well-placed groups, which only serves to consolidate an unequal hierarchical structure.

Within each CoBa, the composition of the board is significant. The members of each board have very close family ties or blood ties *(fati-drà).* They are either father and son, brothers, father-in-law and son-in-law or brothers-in-law. The chairmanship and treasury are held by the family. If there is no one in the family who can read and write, it hands over the secretariat to other literate families.

On the coast, because of the latent competition between the *Temitongoa* and *Temilahehe* territorial groups, there is an almost perfect alternation of power at executive level. If a *Temilahehe* is elected president, the board member must be from the same territorial group as the president and the two territorial groups do not criticise each other to avoid confrontation. By way of illustration, here is the case of *Ambolisogno.* This is one of the 5 *Fokontany* that make up the *Mizakamasy* CoBa. During the first contract, the inhabitants of the *Ambolisogno Fokontany* were full members of the *Mizakamasy* CoBa. The *Ambolisogno Fokontany* was expelled after the extension because they were clearing the park for cultivation. Because the *Ambolisogno Fokontany* is one of the other non-native *Fokontany*, the other members of CoBa *Temitongoa* and *Vondrone* blame them for their habit of clearing land. In fact, their land is on the edge of the park and their fields were bordered by the *Tsimanampesotse* park when it was being extended.

As the *Fokontany* of *Ambolisogno* is a member of CoBa *Mizakamasy*, MNP and WWF obliged the 4 *Fokontany*[84] members to expel the people of *Ambolisogno* from CoBa. MNP continued to take legal action against the *Ambolisogno* clearers while we were in the field. The people of *Ambolisogno* do not have the support of the other communities belonging to CoBa because of their origin (they are former immigrants or *valovotaka.* They do not always have landlord or *tompon-tany* status).

In the interior of the *Mahafale* plateau, in the *Mitsinjo Taranake Magnasoa Tane* CoBa in *Marofototse,* only a "*vahiny*" teacher who arrived in 2000 and holds the post of secretary-treasurer is educated (a teacher) and the other key posts in the office are held by the *village's Zanakanga* communities. The latter are notables who have already held or hold social and administrative positions[85] . The composition of the board by election gives no representation, and therefore no power, to young people. However, the teacher is young and dynamic. WWF entrusted all activities to this young teacher. He is also being courted and groomed by the WWF field team to chair the *Marofototse* CoBa. he notables of the *Fokontany* did not appreciate this programme and the teacher *(*of the *Lanivato* caste*)* was murdered two months before the election to preventhim from becoming the head oftheZanakanga. The teacher was murdered by so-called *malaso*700 m from the village of the current president.

[84] Fokontany de Nisoa, Tsiadriona nord, Ambalabe Beavoha and Sakariake.

[85] Fokontany chief and local councillor.

It should be remembered that thecommunity must have exclusive rights over the forests or resources to be managed. The criteria governing membership of the group of qualified forest resource users must be clear. The group must firmly agree on its composition. It is undoubtedly preferable for the membership rules to allow for the rapid growth of members, because for the management of natural resources to be effective, the entire community must adhere to this transfer of management. If the community admits both village residents and all their relatives living in neighbouring villages to the user group, community resources will come under pressure unless a method is devised to match the size of the group or its overall demand with the capacity of the resource system. This restriction does not mean that the community will exclude outsiders, but anyone from outside the community must have the approval of the community management structure in order to access the resources. The management structure can impose conditions of access, enforce their management plan and can demand payments in cash or in kind for access to the resources. No community would invest in natural resource management if people from outside the community had free access to their resources.

We have used these two cases as illustrations, but there are many asymmetries in the management of each CoBa on the *Mahafale* plateau.

The first indicator of representativeness in the composition of the board is the chairperson, who must be a member of a founding group, whatever his or her level of education, and a scholar from outside the community or a founding territorial group must be excluded.

As for the members, the over-representation of an ethnic group does not pose a problem insofar as it reflects the 'sharing' of rights of access to and use of the territory's resources. This is why it is difficult to establish shared management of renewable natural resources, which remains illusory in Madagascar because societies do not operate on a democratic and participatory basis [Blanc-Pamard and Fauroux Emmanuel, 2004].

Within the framework of EP III, the various environmental operators noted in 2005[86] that the main problem revealed during an evaluation of transfers of natural resource management remains the design of the management: its implementation (with the question of the duration of supervision) and the monitoring of the management transfer. Social conflicts and governance problems within the CoBa are often the biggest obstacles to the sustainable management of renewable natural resources, hence the need for socio-organisational support. This observation is still valid for the CoBa in the *Mahafale* landscape.

This practice involves taking into account the issues surrounding resources, rules of access and use, and what Fauroux has called *'micro-local power structures'*. But is it possible to put in place an appropriate management contract for each of the communities that embark on this path, i.e. to make the contract part of the system's logic so that the transfer does not lead to a rupture or reinforcement in the social organisation by redefining the territory, access and use of resources?

There were other anomalies. First of all, participation is only partial because some players, described by Maldidier as *"absent third parties"*, whose role is important, come voluntarily or involuntarily from outside the CoBa. These are agro-pastoralists, loggers, traders, etc. who prefer to pursue their activities in a legal manner and look after their own interests.

[86] In PE III report, 2005

The rural commune is also absent insofar as the contract is bipartite between the grassroots community and the ecology, environment and forestry department. The rural commune is not a partner, as it is in the Gelose management transfer. Furthermore, the GCF management transfer is based on a contract between the forestry administration and the grassroots community. Yet it is not only between these two partners in the "local management of renewable natural resources" that conflicts arise, but also between members of the same community (belonging to different lineages), between newcomers and natives, between neighbouring rural communities and even between support organisations and service providers adopting different systems.

4.1.5.2 Gender approach in the grassroots community

There has been an improvement in the CoBa statute compared with the previous statute. The article on the term of office of the officers is the one most affected by this improvement. The officers are elected by the members of the CoBa for a renewable two-year term, whereas the old statutes stipulated that the officers performed their duties until the end of the management contract, i.e. for 3 years. The members noted the various kinds of abuses committed by the management committees during the first contract and reduced the term of office of the executive committee.

Those newly managing their land have also followed the experience of previous transfers. They have limited the mandate of an office to two (2) years and if it is renewed, it will be responsible for renewing the second contract, which will last 10 years.

To uncover other hidden problems linked to the existing structure of CoBa, we will try to find out more about the management style adopted by the committee and who makes up the offices?

Thus, studying the composition of the boards helps us to understand whether the transfer of management leads to an investment by outsiders in the village community or reinforces the existing power structure that the members have denounced as abusive. In both cases, the system is asymmetrical and tensions arise between these powers: foreigners versus indigenous groups, strong groups versus socially less well-placed groups, which only serves to consolidate an unequal hierarchical structure.

Within each CoBa, the composition of the board is significant. The members of each board have very close family ties or blood ties *(fati-drà)*. They are either father and son, brothers, father-in-law and son-in-law or brothers-in-law. The chairmanship and treasury are held by the family. If there is no one in the family who can read and write, it hands over the secretariat to other literate families.

On the coast, because of the latent competition between the *Temitongoa* and *Temilahehe* territorial groups, there is an almost perfect alternation of power at board level. If a *Temilahehe* is elected president, the board member must be from the same territorial group as the president and the two territorial groups do not criticise each other to avoid confrontation. By way of illustration, here is the case of *Ambolisogno.* This is one of the 5 *Fokontany* that make up the *Mizakamasy* CoBa. During the first contract, the inhabitants of the *Ambolisogno Fokontany* were full members of the *Mizakamasy* CoBa. The *Ambolisogno Fokontany* was expelled after the extension because they were clearing the park for cultivation. Because the *Ambolisogno Fokontany* is one of the other non-native *Fokontany*, the other members of CoBa *Temitongoa* and *Vondrone* blame them for their habit of clearing land. In fact, their land is on the edge of the park and their fields were bordered by the *Tsimanampesotse* park when it was being extended.

As the *Fokontany* of *Ambolisogno* is a member of CoBa *Mizakamasy*, MNP and WWF obliged the 4 *Fokontany*[87] members to expel the people of *Ambolisogno* from CoBa. MNP continued to take legal action against the *Ambolisogno* clearers while we were in the field. The people of *Ambolisogno* do not have the support of the other communities belonging to CoBa because of their origin (they are former immigrants or *valovotaka.* They do not always have landlord or *tompon-tany* status).

In the interior of the *Mahafale* plateau, in the *Mitsinjo Taranake Magnasoa Tane* CoBa in *Marofototse,* only a "*vahiny*" teacher who arrived in 2000 and holds the post of secretary-treasurer is educated (a teacher) and the other key posts in the office are held by the *village's* Zanakanga communities. The latter are notables who have already held or hold social and administrative positions[88] . The composition of the board by election gives no representation, and therefore no power, to young people. However, the teacher is young and dynamic. WWF entrusted all activities to this young teacher. He is also being courted and groomed by the WWF field team to chair the *Marofototse* CoBa. The notables of the *Fokontany* did not appreciate this programme and the teacher *(*of the *Lanivato* caste*)* was murdered two months before the election to prevent him from becoming head of the *Zanakanga.* The teacher was murdered by so-called *malaso* 700 m from the village of the current president.

It must be remembered that the community must have exclusive rights over the forests or resources to be managed. The criteria governing membership of the group of qualified users of forest resources must be clear. The group must firmly agree on its composition. It is undoubtedly preferable for the membership rules to allow for the rapid growth of members, because for the management of natural resources to be effective, the whole community must adhere to this transfer of management. If the community admits both village residents and all their relatives living in neighbouring villages to the user group, community resources will come under pressure unless a method is devised to match the size of the group or its overall demand with the capacity of the resource system. This restriction does not mean that the community will exclude outsiders, but anyone from outside the community must have the approval of the community management structure in order to access the resources. The management structure can impose conditions of access, enforce their management plan and can demand payments in cash or in kind for access to the resources. No community would invest in natural resource management if people from outside the community had free access to their resources.

We have used these two cases as illustrations, but there are many asymmetries in the management of each CoBa on the *Mahafale* plateau.

The first indicator of representativeness in the composition of the board is the chairperson, who must be a member of a founding group, whatever his or her level of education, and a scholar from outside the community or a founding territorial group must be excluded.

As for the members, the over-representation of an ethnic group does not pose a problem insofar as it reflects the 'sharing' of rights of access to and use of the territory's resources. This is why it is difficult to establish shared management of renewable natural resources, which remains

[87] Fokontany de Nisoa, Tsiadriona nord, Ambalabe Beavoha and Sakariake.
[88] Fokontany chief and local councillor.

illusory in Madagascar because societies do not operate on a democratic and participatory basis [Blanc-Pamard and Fauroux Emmanuel, 2004].

Within the framework of EP III, the various environmental operators noted in 2005[89] that the main problem revealed during an evaluation of transfers of natural resource management remains the design of the management: its implementation (with the question of the duration of supervision) and the monitoring of the management transfer. Social conflicts and governance problems within the CoBa are often the biggest obstacles to the sustainable management of renewable natural resources, hence the need for socio-organisational support. This observation is still valid for the CoBa in the *Mahafale* landscape.

This practice involves taking into account the issues surrounding resources, rules of access and use, and what Fauroux has called *'micro-local power structures'.* But is it possible to put in place an appropriate management contract for each of the communities that embark on this path, i.e. to make the contract part of the system's logic so that the transfer does not lead to a rupture or reinforcement in the social organisation by redefining the territory, access and use of resources?

There were other anomalies. First of all, participation is only partial because some players, described by Maldidier as *"absent third parties",* whose role is important, come voluntarily or involuntarily from outside the CoBa. These are agro-pastoralists, loggers, traders, etc. who prefer to pursue their activities in a legal manner and look after their own interests.

The rural commune is also absent insofar as the contract is bipartite between the grassroots community and the ecology, environment and forestry department. The rural commune is not a partner, as it is in the Gelose management transfer. Furthermore, the GCF management transfer is based on a contract between the forestry administration and the grassroots community. Yet it is not only between these two partners in the "local management of renewable natural resources" that conflicts arise, but also between members of the same community (belonging to different lineages), between newcomers and natives, between neighbouring rural communities and even between support organisations and service providers adopting different systems.

4.1.5.3 Technical capacity building

One of the measures to be taken is to provide ongoing leadership and supervision of the CoBa, or at least during the first year of implementation for new TGRNs and during the first financial year of the contract for renewed TGRNs.

It has to be said that, despite their motivation and goodwill, the CoBa, most of whom are illiterate, very often lack the drive to carry out their activities.

The fundamentalproblem with the *Mahafale* plateau lies in the lack of self-confidence of the CoBa, as they don't always know what they should be doing.

A few concrete examples support this analysis. In almost all the TGRNs, the achievements of the CoBa's respective annual work plan (AWP) are limited to patrols in the forests and reforestation. This may be a coincidence, but the role of patrols has been well explained to the CoBa.

[89] In PE III report, 2005

For other activities, such as raising community awareness of the TGRN or of management instruments, such as reporting to the forestry administration and/or Communes, such as setting up a firewall, etc., no training has been provided. In the case of awareness-raising, for example, the members of the *Itomboina* and *Zamasy* CoBa have admitted that they do not master either the TGRN or the provisions of the various management instruments in order to raise awareness. Similarly, when it came to reporting to the Forest Administration and/or the local Commune, the majority of CoBa said that they were either unaware of the obligation to report, or that they were unaware of the provisions or content to be included in a report. As for the firewall, following the example of the *Ekelelahy* CoBa which, during the evaluation of its TGRN, did not hesitate to ask the forestry agents and those of WWF to explain to them how to build a firewall, the other CoBas are surely waiting for training or at least technical instructions before they can do so.

Community management of natural forests must be based on the principle of sustainability. The exploitation of these natural resources must be carefully considered. We need to think about the regeneration of these resources so that future productivity does not decline. However, there is no precise definition of the sustainable management of natural forests, and no sustainable management system that has already "proved its worth" in Madagascar. The relevant question to ask in this situation is what is the optimal link between the transfer of management and the type of management plan to require prior to the transfer. Should the forest administration require a detailed management plan from the CoBa before granting a management transfer, or would the CoBa's commitment to the principle of sustainability be sufficient to obtain this agreement? In any case, the regulations governing logging must respect the limits of the system and be environmentally prudent enough to allow for a margin of error.

Drawing up a detailed management plan is highly technical, and the CoBa have a very low level of education. They have encountered technical problems because the implementation of this new management policy requires capacity-building for members in :

- the installation of the *"Aro afo"* firewall;
- patrol technique ;
- Techniques for setting up a nursery;
- mastery of harvesting and reforestation techniques.

There are not enough technical trainers to train members. Only one officer per district is responsible for running the forestry administration in his area. This officer is called the chef de cantonnement. This inadequacy prevents the new management system from being properly mastered. The success of this management and the promotion of a well-protected environment far from any form of degradation is a necessary condition for the sustainable development of a given region or country. The technicians who provide the training needed to implement the transfer can pose problems, as they are not familiar with the current situation, which means that the techniques provided are unsuitable.

There is never any technical capacity building provided by the cantonment chief to the CoBa who manage the forest resources. "*...we do what we can and practice what we know to protect the forest even though we have never received any training or techniques to protect our forest"* [Focus group in *Antanindranto*].

4.1.5.4 Land disputes on the increase

The local people know the boundaries of their land and of the transferred forest areas. The administrative bodies do not have a tool for locating these areas. As a result, the[90] aspects of land tenure security under the Gelose Act have not yet been fully implemented during the first management transfer contract. Moreover, the land office that will handle this aspect does not yet exist in the *Mahafale* landscape.

Local people generally use customary rights to gain access to land, and their practices ignore the law established by the State. Management transfers do not address the land issue. Against this backdrop, there has been a resurgence of open or latent land conflicts in the *Mahafale* plateau.

Figure 4: Conflict over land in relation to the respondent's professional activity

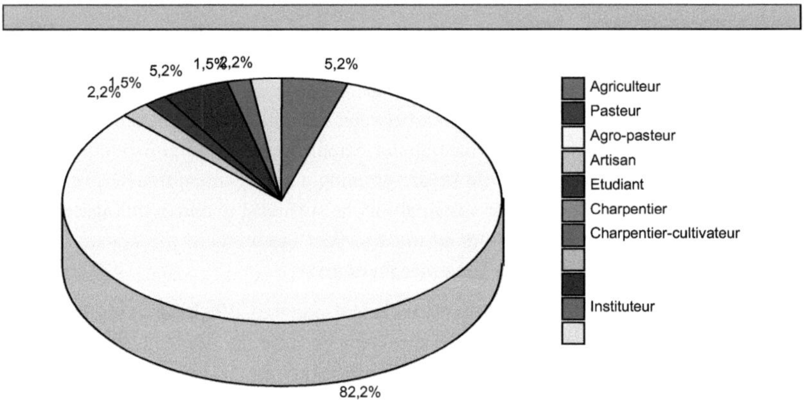

Source: Our own survey, 2014

The communities that have already experienced or are still indirectly affected by this land conflict linked to their professional activities have confirmed its existence at community level. To illustrate this, our survey of CoBa members and non-members in the management transfer zones revealed that 82.2% of the agro-pastoralists surveyed confirmed this, 5.2% of each of the farmers and shopkeepers surveyed had experienced this land conflict and 2.2% of each of the

[90] Relative land tenure security has three distinct dimensions:

- The delimitation of the village terroir provides a territorial basis for the grassroots community responsible for the consequences of the dynamics of land appropriation implemented in the area covered by the management transfer.
- The delimitation of the perimeter carrying renewable resources (e.g. forest) aims to establish the management of renewable natural resources conferred by the management contract. Mapping helps to clarify the commitment to conservation and sustainable use that is the aim of the management transfer.
- Finally, the identification of land uses corresponds to a more detailed level of land organisation, specifying the tenure rights of the members of the basic community (lineages or segments of lineages, individuals).

craftsmen and teachers also testified to the resurgence of land conflict in the *Mahafale* landscape.

Figure 5: Land conflicts among farmers surveyed

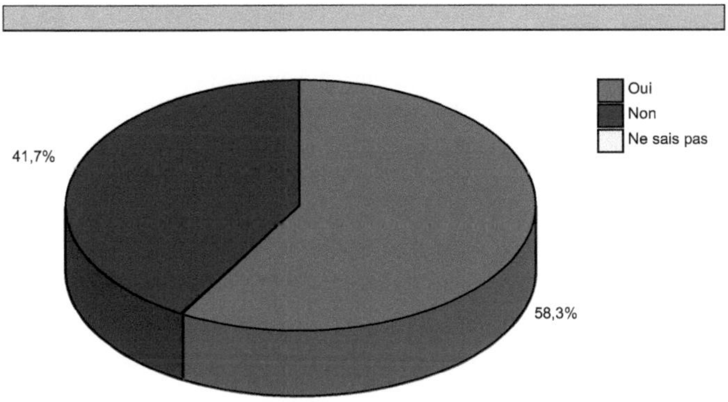

Source: Our own survey, 2014

This figure shows the extent of the conflict over land among farmers: 58.3% of respondents who were members or non-members of the CoBa said they had experienced this conflict over land. This conflict is due both to the extension of the park and to the *Mahafale* tradition of building an 8 x 8 metre tomb to bury a single dead person. Nothing can be done around the tomb as it is taboo or *"faly"*. To overcome this problem, the communities grow crops discreetly inside the park.

4.1.5.5 Grassroots communities and climate change

Madagascar is a country where poverty in rural areas is still endemic. These poor people depend on natural resources and are therefore sensitive to climatic shocks. The considerable impact of the latter can be seen in the field of our research.

Firstly, the *Mahafale* plateau has a sub-arid or semi-arid climate. It is a landscape of plateaux and plains that forms part of the Saharan regions. It is characterised by a long dry season lasting 7 to 9 months. The rainy season is often very irregular and always poor in rainfall (less than 600mm/year)[91] . The rainy season has become very irregular. The planning of activities that are more or less directly conditioned by rainfall is disrupted and has become hazardous. The usual regular and prolonged rains are giving way to torrential rains of short duration, which are difficult to control (storage, watering of crop fields, etc.). In particular, climate change will increase drought in the deep south, but also in the *Mahafale* landscape, making communities more vulnerable.

[91] In Testimonies from Madagascar: "Climatechange and rural lifestyles", WWF 2010

Every time there is a drought, people who are already very poor lose their tools of the trade and therefore their means of subsistence. According to WWF, more than 78% of the total population of the *Mahafale* plateau live below the poverty line. Farming and livestock rearing are the most common activities, although arable land accounts for no more than 15% of the plateau's total surface area. The social and economic life and livelihood of most of the communities still depend to a significant extent on natural resources. This dependence, fostered by a lack of development opportunities, precarious living conditions and poor access to modern technologies (communication, education, electricity, health, etc.), increases the vulnerability of communities when natural resources are vulnerable, and vice versa.

Figure 6: Agricultural production compared to 5 years ago

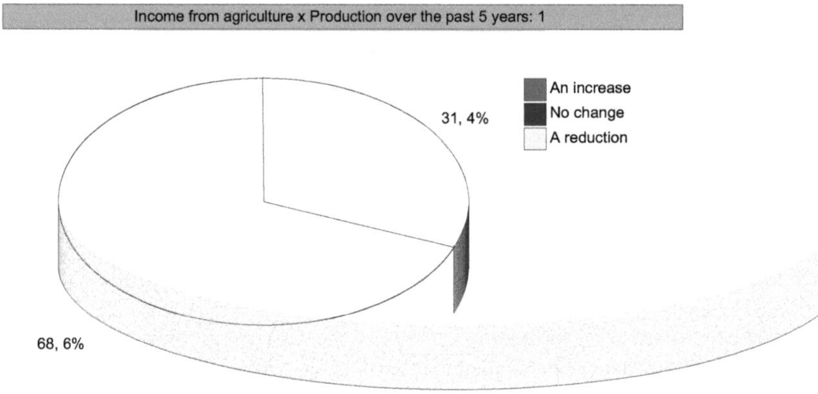

Source: Our own survey, 2014

Figure 6 shows the impact of climate change on the agricultural activities of communities in the *Mahafale* landscape. Compared with 5 years ago, the respondents noted a remarkable drop in their agricultural production. 68.6% of them confirm this. While 34.1% said that their current production is the same as it was 5 years ago. None of the respondents said that their production had increased. This means that both their annual production and their field are deteriorating from year to year. During the focus group we organised, most of them adopted a different food and/or cooking habit during the lean period. Since people no longer have enough produce to support themselves and to sell, they no longer have any money. This situation changes their cooking and/or eating habits: *"The price of cattle and goats has become very low, while the price of other basic necessities has doubled. As a result, we drink 'rohondroho[92] ' every morning instead of coffee."* [Focus group in *Zamasy*].

But this change in the culinary habits of certain communities has no impact on climate change. We will try to explain the action taken by communities in the face of climate change using the figure below.

[92] A tree from the *Mahafale* region whose leaves and bark can be used to make herbal teas.

Figure 7: Action taken in response to climate change

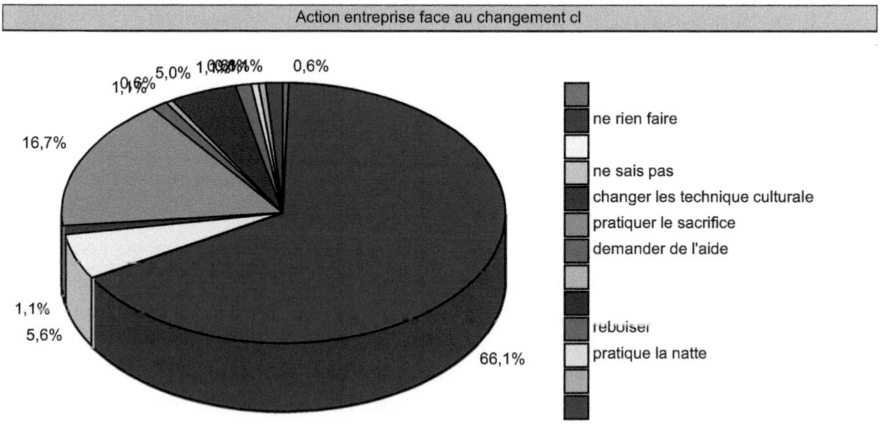

Source: Our own survey, 2014

The 66.1% of people surveyed have done nothing about the drought caused by climate change. Communities are at a dead end and do not know what to do to overcome this natural disaster. On the other hand, 16.7% practice the sacrifice for the year known in Malagasy as *"soron-tao*[93] *"* to ask their gods and ancestors for abundant rain. They believe that by sacrificing an ox, the gods and ancestors can hear them and will remedy the situation. They perform the sacrifice just before the cyclone season and believe that the gods and ancestors will hear their requests. They perform it every year, even though the rains only fall during the cyclone season in Madagascar.

5.6% of respondents have changed the seeds they use. They are trying to use improved seeds. Also, 5.0% of them said they were protecting forests to combat climate change.

Only 1% of them admitted to having changed their cropping technique. They are adopting a system of cultivation known as semi-cover cropping or SCV. To implement the "Improvement of cropping and pastoral systems" component, the WWF, in partnership with AVSF (an operator disseminating the SCV cropping system), has popularised conservation agriculture to help the grassroots communities that manage the natural resources around *Tsimanampesotse* Park to alleviate the food insecurity that is rife on the plateau.

Despite the introduction of this new cultivation system by WWF in the *Mahafale* plateau since 2005, farming has remained very outdated for most farmers, and 95% of plantations do not use any fertiliser, whether mineral or organic[94] . The *Mahafale* communities are very conservative and are inaccessible to all the changes put forward by development promoters.

[93] A ritual performed by the *Mahafale* people every year to ask for abundant rain. This rite is performed between October and December by sacrificing a zebu. The zebu remains at the centre of communication between the living and the lineage ancestors, on whom prosperity ultimately depends. It is always by sacrificing zebus, the wealth par excellence, that the quality of this communication is ensured.

[94] Agricultural Statistics Service, MAEP, 2006

4.1.5.6 Management Transfers accepted but not appropriated by communities

During the focus group we organised, villagers of all ages were surprised. They had noticed that if deforestation increased, there would be nothing left for future generations. As a result, they saw the need for rational management to regulate the rate of pressure on natural resources. Some villagers have even realised that deforestation is one of the causes of climate change. These villagers joined the TGRN in order to ensure the sustainability of the habitats in their areas, which they say are threatened above all by pressure from people outside their areas.

In any case, this general awareness is borne out by the fact that even though the majority of villagers are not members of the CoBa, and even though several villagers (CoBa members or not) did not participate in drawing up the *dina* or the simplified natural resource development and management plan, no villager opposed the establishment of the TGRN. Nor did anyone refuse to comply with the provisions of the *Dina*, and there were no reports of CoBa being called into question as a natural resource manager. However, in the majority of TGRN territories, CoBa members are far from representative.

Even if the TGRN are accepted by the communities of the terroirs around the *Tsimanampesotse* park, they have not appropriated them. In the light of discussions and interviews with villagers from the various terroirs, this lack of appropriation is due mainly to two facts:

- The first relates to awareness. Some villagers claim not to have been made aware of the TGRN, others who have not attended one or more awareness-raising sessions admit not to have understood much, and still others think that the TGRN is a matter reserved for members of the CoBa board. In short, the concept of management transfer escapes the communities, and a misinterpretation of CoBa membership has been perpetuated by the lack of ongoing awareness-raising;
- This lack of appropriation is due again, and perhaps above all, to the absence of ritualisation. This lack of ritualisation was raised many times, especially by the *mpitan-kazomanga* (traditional authorities). It should be noted that the head of the MNP sector based in *Itampolo* also made this remark before our field visit.

One of the primary psychological reactions of village communities is to blame all disasters on a failure to observe traditions. Those who dare to act contrary to tradition are singled out when disaster strikes. The villagers, even if they are convinced of the merits of the TGRN, do not dare get involved for fear of provoking the wrath of the ancestors by transgressing traditions. Their support for the new organisation brought about by the TGRN is conditional on the observance of a ritual that establishes harmony and understanding between the ancestors and the living.

4.1.5.7 Base community and application of the *dina*

In the transfer of natural resource management, the management contract incorporates the *dina* or village convention. Law no. 96-025 on the local management of natural resources, known as the Gelose law, recognises the *dina* and its role in the community. Relations between members of the community are regulated by means of a *dina*. This is approved by the members of the basic committee in accordance with the customary rules governing the community. The text uses the *dina* as the basic law. In this way, local communities define the rules of use that will govern the management of common resources and areas within their territory.

In other words, it is the members of COBA themselves who draw up their own *dina*, but this is done with the approval of the commune and the District, and homologated by the court of first instance so that it has the force of law.

However, the provisions contained in this village agreement must comply with the constitutional, legislative and regulatory provisions in force, as well as with recognised and uncontested practices in the rural commune to which it is attached. The law makes the *dina* subject to the authority of the mayor of the commune to which it is attached.

However, in the *Mahafale* plateau, it would be an illusion to believe that all is well with the application of this *dina*. There have been a few reports of *vonodina* not being applied[95] and of the *dina* itself not being applied. It has been said that this type of situation is due to the way the CoBa are structured: the weight of traditional authority, alliances and family ties, the social status of offenders, etc.

It seems that the application of the *dina* is only called into question in two cases. Firstly, when the CoBa's partners - the supervisory commune and the forestry administration - fail to meet their obligations with regard to the *dina*: distant follow-up, complaints and reports concerning infractions that have been duly noted but not acted upon. Secondly, when the CoBa is set up with no regard for local realities.

The *dina* is modelled by WWF agents on the outline of a standard agreement drawn up by forestry administration technicians at central level.

Figure 8: The community's interest in joining CoBa

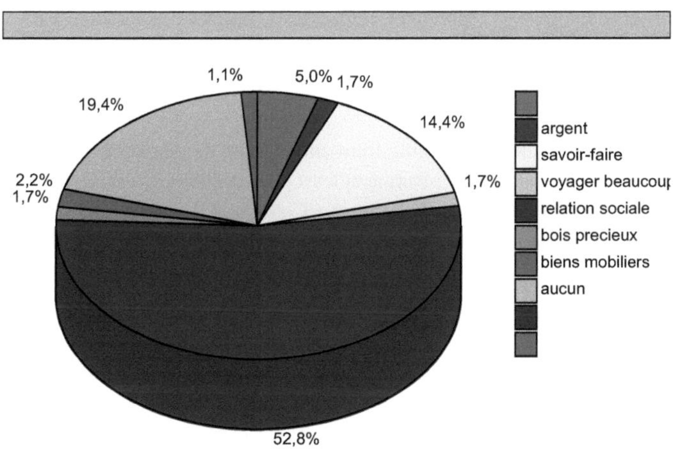

Source: Our own survey, 2014

[95] The *vonodina*, a corollary of the dina, sets out the measures to be taken in the event of an offence or transgression. The *Ray amandreny mpizaka* responsible for its application are called "*mamono dina*".

In this graph, if members are asked about their interests in joining CoBa, 52.8% say they have new social relationships; 19.4% say they have no interests, while 14.4% say they have know-how. 5% chose not to answer the question; 2.2% said they had received movable assets. 1.7% benefited from precious woods or travelled a lot or had money. Finally, 1.1% did not answer this question because the respondents hesitated between having no interest and benefiting from wood.

While this figure reveals the community's interest in joining CoBa, the data received has an impact on the application of the *dina* within CoBa. For most of them, joining CoBa increases their relationship with others within or outside CoBa. As we reported in this chapter, CoBa board members all have a family connection. This situation disrupts the application of the *dina* in their community. Family members who are not in power take advantage of this situation by infringing the established *dina* in the use of forest resources. The case of land clearing by the people of *Ambolisogno* is proof of this: these people are from the same territorial group as most of the members of the *Nisoa* CoBa office. The office has tolerated people who violate village conventions, even if they are caught red-handed. The organisations involved in the conservation of the *Tsimanampesotse* park (MNP and WWF) are forcing the office to take drastic action against the culprits.

The intervention of public officials such as the gendarmerie, the commune and the chef cantonnement forestier weakens the application of the *dina* and they put pressure on the offices to deal with the case amicably without taking into account the *vonodina* in cases of flagrante delicto.

4.1.5.8 Grassroots community and reforestation

All the CoBa on the *Mahafale* plateau already have a management plan. To make the most of them, annual reforestation is compulsory. Moreover, reforestation has been included in a PTA as one of the CoBa's priority activities with the aim of conserving and sustainably managing the transferred forests. They will have to reforest both seedlings that they can use for their use rights and for an income-generating activity.

Reforesting requires money. The price of a seedling from nurseries on the *Mahafale* plateau is 450 *ariary*. Each CoBa sets itself the target of planting at least 300 seedlings a year. Given that the membership fee is the only source of income for the Coba, and that members are not motivated to pay it every month, each CoBa's fund is insufficient for the monthly[96] patrols and the purchase of seedlings.

Looking at the CoBa reports, we found that most CoBa's have replanted twice since 2011. Members have transplanted wild saplings from the hard core to the user rights area. Most of these saplings are *katrafay* and *kapaipoty*. These two types of tree are the ones most used by the communities in their daily lives. The CoBa of the *Mahafale* landscape do not carry out much reforestation, even though this is a key element in the sustainable use of forest resources.

However, because of the drought on the *Mahafale* plateau, the hunger gap lasts a very long time and the grassroots communities are becoming much poorer. As a result, the support project is

[96] Each time the CoBa conducts a patrol, the cost is estimated at 40,000 ariary on average. The designated members patrol for 4 days. The closest distance from the village to the transferred forests is 15 km.

trying to find alternatives, firstly to ensure the long-term management of these natural resources, and secondly to improve the CoBa's standard of living.

The COGESFOR project, co-funded by WWF and AFD, has therefore decided to introduce the *moringa olifeira (ananambo)* reforestation project as an alternative that will help the CoBa to find ways of combating poverty.

The CoBa have agreed to adopt the reforestation of this species of *moringa olifeira* because it is adapted to the climate of the *Mahafale* plateau: it is resistant to the dry, arid climate. What's more, this tree can be used for shade in the future, but growing this plant will also help the CoBa to improve their health by using water treated with *moringa* pods. The plant's leaves can be used as bedding and will provide the CoBa families with a complete source of nutrition.

In addition, there is a guaranteed outlet, as the company Philéol has signed an agreement to buy and sell *moringa* products from CoBa member growers.

However, after 2 years, the project failed. Only three CoBa are able to produce *moringa* leaves: *Ekelelahy*, *Ampitanake* and *Vorojà*. The first two (*Ekelelahy* and *Ampitanake*) are on the banks of the *Ilinta* river and their crops are protected from drought. *Vorojà* is on the edge of the *Tsimanampesotse* park. A Madagascan association based in Switzerland has signed a contract with these three CoBa.

4.1.5.9 Grassroots communities faced with poverty

According to the people most involved in the movements for the management of their forest resources, the poor do not mobilise voluntarily for their own; it is rather money and/or the '*dina*' that play a major mobilising role. However, vulnerable groups in the community, such as young people, women and the poor, have taken on the thankless work, despite the situation described above and, above all, the monopolisation of power by the leaders.

The poor make up the majority of the population of the *Mahafale* landscape, but they don't have much responsibility. Female poverty and the problem of youth unemployment (a rush to the sapphire mining areas and the rice fields of *Morondava*) accentuate this phenomenon. For the poorest section of the population, who are the most active in work requiring physical effort, the WFP's labour-based projects are seen as a source of income.

What's more, as they don't see any tangible results very quickly, the poor quickly become discouraged. They prefer to focus on their day-to-day concerns, especially as the patrol lasts four or five days.

The failure of communication and the strategy of the leaders have contributed to the exclusion of the poor from taking responsibility or even simply participating in the villages. Whether they consent or not, their ignorance of what is happening in the *Mahafale* landscape and even in the communes and neighbourhoods is an indicator of how excluded they are.

4.1.5.9.1 Incapacity of the poor

We believe that the following two elements shape citizens' initiative or other forms of commitment to community action: their need to develop as people and their individual capacity. This is why we have chosen this hypothesis.

Poverty erases human dignity and civic sense. The poor, as mere doers without the slightest initiative, cannot hope for any satisfactory social development or position, or any improvement in their material and living conditions. Poverty, whether intellectual, cultural or economic, thus creates a feeling of frustration or dissatisfaction and can easily lead to a lack of interest in community responsibilities, as the table below shows.

Table 8: Level of participation by level of education

Niveau d'instruction	oui	non	TOTAL
aucun	48,4%	51,6%	**100%**
T1	60,0%	40,0%	**100%**
T2	60,0%	40,0%	**100%**
T3	60,0%	40,0%	**100%**
T4	57,1%	42,9%	**100%**
T5	60,0%	40,0%	**100%**
6	75,0%	25,0%	**100%**
5	100%	0,0%	**100%**
4	0,0%	100%	**100%**
3	40,0%	60,0%	**100%**
Seconde	33,3%	66,7%	**100%**
	100%	0,0%	**100%**
terminale	50,0%	50,0%	**100%**
	0,0%	0,0%	**0,0%**
	0,0%	0,0%	**0,0%**
TOTAL	**51,1%**	**48,9%**	**100%**

Source: Our own survey, 2014

According to this table, in general, the people surveyed who had completed primary school were the most active members of CoBa, not to mention those who had completed 6th and 5th grade, who acted as facilitators in villages where forest resource management had been transferred. On the other hand, those who have never been to school have little motivation to join CoBa (51.6% compared with 48.4%). People who have already passed the 4th grade should not be overlooked, as they do not participate much in CoBa life because of their activities. In fact, most of the people who have a secondary education have as their professional activities the collection of local products or the resale of generic medicines. They have no time to devote to CoBa activities.

Only one member of the CoBa in *Zamasy* (secretary-treasurer) has the level of first class, hence the percentage of 100%. It should be pointed out that this person always represents CoBa with the president at every meeting or training session.

Table 9: Degree of adherence to CoBa in relation to professional activities

	Agriculteur	Pasteur	Agro-pasteur	Artisan	Etudiant	Charpentier	Charpentier-cultivateur	Agent vulgarisateur de sa	ant	Instituteur		TOTAL
oui	6,5%	0,0%	79,3%	2,2%	1,1%	0,0%	1,1%	1,1%	6,5%	0,0%	2,2%	**100%**
non	6,8%	0,0%	83,0%	1,1%	1,1%	0,0%	0,0%	0,0%	4,5%	2,3%	1,1%	**100%**
TOTAL	**6,7%**	**0,0%**	**81,1%**	**1,7%**	**1,1%**	**0,0%**	**0,6%**	**0,6%**	**5,6%**	**1,1%**	**1,7%**	**100%**

Source: Our own survey, 2014

According to this table, communities in the Mahafale landscape have 11 main activities, namely agriculture, livestock farming, agriculture and livestock farming, handicrafts, carpentry, schooling, agriculture and carpentry, health extension, trade, fishing and education. In the Mahafale zone, wealth is measured by the number of head of cattle. Farmers and agropastoralists are not motivated to join CoBa because their activities are directly linked to the forest. They feel excluded from the benefits of using their ancestors' forest. Pastoralists are involved in transhumance and do not have much time for CoBa meetings.

Retailers are the most motivated to join CoBa. They position themselves as sponsors of the patrol officers.

Whatever their economic activities, the rich or those on average incomes and all the current or former leaders in the communes are not very involved in the life of the CoBa which manages the forest resources.

The wealthy people surveyed are not involved in the management of forest resources through management transfer because they do not feel very close to the communities because of where they live and their daily activities. They often travel to the regional capital. On the other hand, the moderately wealthy inhabitants are merely participants, as they were very busy with their jobs, being for the most part resellers of generic medicines or small collectors of local products in the *Mahafale* plateau. Otherwise, the poor made up most of the simple participants; they did not have much responsibility but only told the committees about their aspirations. However, most of the poor people surveyed did not take part in the process of setting up the CoBa, mainly because of their daily occupations.

78% of the population surveyed are poor. This classification was assessed on the basis of monthly income, if not livestock numbers: most of those surveyed do not even earn 50,000 *Ariary* per month. Yet they belong to the active population of the *Mahafale* plateau.

What's more, the survival instinct in humans, as in all animals, means that they are always able to find food even in extreme poverty, but they can't do anything else. So the poorer you are, the more your functional capacity diminishes. We felt that precise knowledge of people's standard of living was important in order to understand the conditions under which people in the *Mahafale* landscape can participate in the conservation of forest resources and sustainable development.

Finally, poverty can become a pretext for avoiding duties towards the community, as it is often difficult to assess the standard of living of each inhabitant. However, according to what we explained earlier, the participation observed in the case of the constitution of a management transfer of renewable natural resources in the *Mahafale* landscape corresponds to the type of participation provoked.

The particularity of the *Mahafale* landscape is that it is home to a large number of illiterate people. We have already noted this situation in the monographic section. Among the illiterate people surveyed, 48.4% were members of the CoBa, compared with 51.6% who were not motivated to join. However, 75% of those who had reached 6th grade had joined CoBa, compared with 25% who had not.

4.1.5.9.2 Daily concerns predominate

The management of forest resources and the establishment of a structure to ensure this management have taken into account the general interest. Community members aspire to short-term benefits that have a direct impact on their survival needs. Indeed, the poor of the *Mahafale* plateau see only one priority: to fill their immediate needs.

The fact that 78% of the population are very poor and are more concerned with their daily lives than with the life of their community explains their lack of interest in sustainable community development. This is especially true of those who have precarious jobs or who commute back and forth between the *Mahafale* plateau and the sapphire mining areas, if not *Morondava*. This phenomenon is in line with Maslow's categorisation in his pyramid of human needs, in which he asserted that the poor have not yet gone beyond the first phase of needs, i.e. physical needs.

Figure 9: Pyramid of basic needs

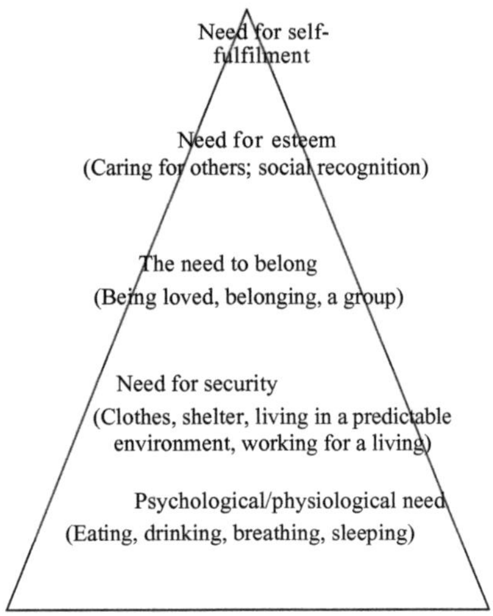

Source: Maslow, A - H, 1987

These physical needs correspond to the need for security shown in the graph above. However, according to the corrected version of this pyramid, the needs of the poor can be classified into two phases, namely the physiological need (to stay alive): this mainly concerns food and rest or recuperation that provides strength and minimum comfort. The second phase is the physical need (to survive), which concerns both health and safety, i.e. avoiding damage or other harm. So the poor think in terms of survival needs, immediate needs.

According to CoBa leaders, the poor only participate if they receive money in return, especially for patrolling. But it seems that money is also a priority, even for those who have joined the management committee. Initially, they had hoped to have access to money and budget management, but when they realised that this was not the case, they scattered. People don't participate, even if it's to improve their village, when they don't earn any money because *"no interest, no action"*. Money is very important for survival, and its absence can block the fulfilment of citizenship duties.

In conclusion, if the sector is to perform well, environmental governance needs to be reformed. This reform must focus on the legal framework for environmental management. It must be simplified and applied. The roles of the various institutions must be redefined, and the participation of civil society in the decision-making process must be improved. This reform must lead to a significant increase in the sector's revenues and an improvement in the ratio of tax revenues to non-tax revenues. Finally, this reform must better integrate and give greater visibility to climate change in the institutional and political landscape.

As far as community participation in conservation is concerned, it can be said that although the management transfer has been accepted, it has not yet been appropriated by the communities themselves. Their participation in conservation is therefore limited because the communities have problems with the implementation of this management transfer. Their main problem is summed up by the sentences uttered on the TV Plus Madagascar news programme on 19 April 2012, saying: *"the VOIs have problems managing their resources because they do not benefit directly from donor funding. In reality, they have no financial autonomy. And because of this, the VOIs are not motivated to take responsibility for the illegal exploitation of their resources. In fact, the management of natural resources is not entirely transferred to the grassroots communities or VOIs. In other words, the pressure exerted by the authorities still persists, which is why the grassroots communities are discouraged from assuming their responsibilities.*

Apart from this problem of financial autonomy, land tenure problems that threaten community cohesion, problems with forest resource management techniques, illiteracy, limited user rights and climatic shocks weaken the communities' motivation to participate in conservation.

CONCLUSION ON THE FIRST HYPOTHESIS

The poor of the *Mahafale* plateau do not care about the projects presented by their leaders and to be drawn up in a priority development programme for the Communes. The aspiration of the poor is really to satisfy their basic needs for survival without the assurance of a minimum of well-being and without the possibility of self-fulfilment. What's more, when the CoBa management committee was set up, the poor were not among those who immediately joined it, as they had neither the patience nor the will, let alone the ability, to do so.

The poor are only motivated to participate if they receive money in return. Apparently, money is also a priority for those who have joined the management committee. Initially, they thought they were going to directly manage an operating budget, but they quickly dispersed when they realised that this was not the case.

Poor people prioritise their day-to-day problems and do not participate, even if it is to improve their village when they do not earn money. The latter is very important for survival, and its absence can block the fulfilment of citizenship duties.

To repeat our first hypothesis: "the level of participation of the grassroots community depends on its level of education, and non-participants at the lower end of the scale do not want to get involved in matters they consider obscure and dangerous".

We can conclude that all our analysis seems to confirm this initial hypothesis. Community members with an average level of education are the most motivated to participate in the conservation of natural resources. They join the CoBa because they want to dominate the community. To gain a social status that can stand up to the rich dominating the villages, they join NGOs. They try to impose themselves on the members of the entire community by concealing information about the management transfer and, above all, about CoBa.

On the other hand, the illiterate and the poorest people do not want to join CoBa. They know little about both the management transfer and the CoBa itself. As a reason for not participating in the conservation and life of the association, they put forward the lack of time to devote to taking part in general assembly meetings. In truth, it is not time that they lack, but rather their conviction and dedication.

As a result, they are unaware of how the situation is evolving, and cannot assume their responsibilities as citizens. In fact, they are informed neither of their right to use these forest resources nor of their duty to protect and manage them. As far as they are concerned, the forests have already been sold to the WWF, the members of the CoBa are merely guardians and prison awaits them in the event of any offences committed in the forest.

In addition, some illiterate people join CoBa in the hope of benefiting directly from the funding that the SGP/Tany *Meva* advocates for communities managing renewable natural resources. They believe that this financial aid will solve their financial problems.

Finally, intellectual poverty creates a feeling of frustration or dissatisfaction and can easily lead to a lack of interest in community responsibilities, particularly the conservation of forest resources.

4.2 Approaches recommended by donors in the process of transferring natural resource management

Today, the management of natural resources is one of the most important environmental concerns. This is why most countries and governments are integrating the environmental dimension into their development policies. After setting up 6,000,000 ha of protected areas, President *RAJAONARIMAMPIANINA Hery* Martial, in his speech in Sydney in 2014, announced that existing protected areas would be tripled. This announcement means that throughout Madagascar, the total surface area of protected areas will increase by 18,000,000 ha.

In this section, we describe the approaches required by donors in terms of community-based natural resource management and the decentralisation of natural resource management in Madagascar.

Before embarking on these steps, it is important to realise that there are two types of model to be considered for the management of natural resources outside protected areas. These models are the community-based management model and the management model with economic incentives. We will focus on the first model because it corresponds to our case.

Historically, the Malagasy have already managed renewable natural resources since the time of the Malagasy royalty and during colonisation up to the present day [MONTAGNE Pierre, RAZANAMAHARO Zo, COOKE Andrew, 2007].

At the beginning of colonisation, the colonists employed a policy of repression and exclusion of the local population from the local management of resources. During this period, they used regulations, the employment of "special personnel" responsible for noting breaches of these regulations and fining the guilty, and the application of penalties through the courts to those guilty of offences.

As a result, this centralised and repressive natural resource management policy failed. Later, donors and the State refocused their new analyses on deforestation and the degradation of natural environments. Both parties unanimously agreed to adopt decentralised management of renewable natural resources. The gradual recognition of local populations gave rise to a national policy of decentralising the management of renewable natural resources in the mid-1990s, with the aim of reclaiming the legitimacy of local resource management from local grassroots communities.

This type of management is called GELOSE or Gestion Locale Securisée. It is an asset management method developed in 1996 to give the people of a village back control over the ecological conditions in which they live.

Since the beginning of the 1990s and the emergence of the 3rd Republic in 1992, Madagascar has been emerging very slowly from a long period of Soviet-style politics led by President *Ratsiraka*[97] , which contributed to the deterioration of the country's infrastructure and living standards. Under the control of donors such as the World Bank and the IMF, various approaches were applied one after the other to rapidly develop Madagascar. These approaches are: the integrated approach, then the concerted approach, then the participatory approach and finally the agar approach. The latter is currently considered to be the only approach that corresponds to the situation in poor countries, as it gives local people the chance to express themselves and negotiate, i.e. it is up to the people to say and imagine what they want for themselves. What kind of development do they want for their children? And all the projects start by helping them to think about their future and their plans for society.

However, we need to ask ourselves whether agar is the result of thinking that says that it is humans who are destroying biodiversity. This question arose after the implementation of the Agar law, which tends to chase people away because they have free access to biodiversity, whereas we know full well that it is impossible to conserve this biodiversity without them.

The technical and financial partners have recommended the following approaches:

First phase:

- at commune level: the grassroots community must declare to the existence of a natural resource that they wish to protect. After , the commune givesits favourable opinion. Next, the community formulates its request for a management transfer, including a list of the resources to be transferred. On receiving the request, the commune sets up an ad hoc committee to investigate the community, give its opinion and thenassess the community's capacity to manage the resources. Finally, the local councillor will submit a request to the District for this management transfer.

Second phase:

- at District level: the District will notify the Ministry of the Environment and Ecology or its regional office of the existence of the request for transfer of natural resource management. The regional directorate for the environment and ecology will give its opinion: if the regional

[97] Close its borders to external trade.

directorate accepts, it will notify both the District and the commune; in the event of refusal, the regional directorate will send a refusal notice including the reason for its objections to the District, which will in turn notify the commune of its refusal.

-at local level :

On receipt of this refusal notice, the municipality will inform the local authority of its decision.

If itaccepts, it will notify the community and, as a result, the grassroots community launches the process of securing relative land tenure by appointing a mediator and selecting consultancy firms to prepare the "securing relative land tenure" dossier[98].Then the community negotiates with the stakeholders to find out how motivated everyone isto manage natural resources.

It then draws up the minutes of these negotiations, and with it the mediator and the consultancy firms carry out studies on the implementation of the management transfer.

Finally, the mediator will draw up the draft management transfer contract (including the management tools).

Third phase:

The grassroots community will send the commune the draft contract. The municipality, through the mayor, the grassroots community and the regional environment directorate will then sign the contract.

The Ministry of the Environment and Ecology or its regional representative will decide on the management transferand will publish the existence of this managementtransfer in the Official Journal. It then approves the management transfer.The municipality registers this approval. As a result, the transfer of management of thebase community will come into effect.

Sophie Godefroit, in her book entitled *La restitution du droit à la parole*thought that la Gélose was an extremely original project, whereas in our view it was modelled on and inspired by colonial policyon environmental management.n fact, at that time,the colonistsemployed a policy ofrepression and exclusion of the populations of the local management of resources. The means they used were regulations, the employment ofspecial staff responsible for noting breaches of these regulations and fining the guilty, and the application of legal penalties to offenders.

he management and use of forest resources during the colonial period was governed by regulations. Special staff were responsible for applying these regulations. In the case of Gélose, the "*dinan'ny tontolo iainana*" replaces the regulations of the colonial period and the special staff is replaced by the grassroots community.The latter plays the role of *andrimaso*

[98] Decree No. 98-610 of 13 August 1998, regulating the terms and conditions for the implementation of Relative Land Tenure Security, in application of Law No. 90-012 of 6 June 1997 amending and supplementing Law No. 90-033 of 21 October 1990 on the Environment Charter, (J.O.R.M., No. 2035, 24.12.90, p. 2540):
Article 1 stipulates that, in application of Law No. 96-025 of September 1996 on the local management of renewable natural resources, the Sécurisation Foncière Relative, hereinafter referred to as SFR, is defined as a procedure consisting of the overall demarcation of the terroir of a local grassroots community benefiting from renewable natural resources, as well as the recording of the occupations included in the terroir.

(vigilance) and punishes the guilty party by means of the habits and customs on plac: the*tsara ambany kily* (traditional court) followed by a *vonodina* or a fine.

The Gelose law does not tolerate free access to biodiversity because it encourages competition between communities. Donors have noted that whoever reaches the forest resources most frequently will ultimately be the best served. So competition has arisen, and it needs to be broken up in order to plan a kind of project: a project built by society itself.

Access to forest resources has been regulated by means of a management transfer, and it is left to the grassroots community to record breaches of this regulation, which is none other than the *dina*, and to penalise the guilty party through local customs and practices: the traditional court and the application of the *vonodina* for those guilty of breaches.

The technical and financial partners thought that the aim of the ban on free access was to restore the rights of nature and the rights over the environment to the population as a whole. However, the application of the Gelose law in Madagascar is sometimes distorted to the benefit of certain individuals.

The ultimate aim of the management transfer around the park is to resolve the anthropic pressure on the forest ecosystems around the *Tsimanampesotse* salt lake. The lake is 15m long and 1m deep. This is to prevent it drying out, as it is a resting place for migratory birds, more specifically the pink flamingo. Central government, through the MNP and with the support of donors, has decided to extend the park. This implies that the pink flamingo is more valuable than the *Mahafale* communities, otherwise they have protected a major interest (mining resources) in the area.

As a result, the technical and financial partners have involved the communities surrounding the park in protecting and managing their forests in a sustainable way through a management transfer. The involvement of the communities urgently requires a number of conditions to be met, such as:

- the management rules must be drawn up and accepted by the members of the community. Theymust be publicly accepted, officially approved bythe local authorities, the community's administrative managers and endorsed by the customary authorities [*Olobe, Mpisoro, Mpitan-kazomanga, Ampanjaka, Tangalamena,* etc.] at a ritualisation ceremony. Once this has been done, acceptance can take the form of a charter or fundamental rules, which are essentially considered as guidelines, to prevent community management from being transformed into open access;
- the community mustit be legally constituted with common interests, its memberses must be culturally homogeneous. These conditions facilitate the management of any internal conflicts that may arise, on the one hand, and the identification ofcommon objectives during the implementation of management, on the other. These conditions also make it easier to capitalise on traditional knowledge and know-how when implementing community management;
- the presence of economic incentives for the community, a condition that must be met if community managementisto function effectively. In most cases, the first economic incentive offered to the community is the one that will improve the income of member households. Indeed, even if bequeathing something for future generations is important,

a poor community that is trying to survive and find enough to eat from day to day will certainly not have the luxury of thinking about the future of future generations when it has none of its own.

Chapter Five: Conditions for the participation of grassroots communities in the protection of natural resources

Democracy is considered to be the most favourable structure for citizen participation. Madagascar has been committed to this system for several years. At present, it is important to know whether the political structure, the culture and the structure of *Mahafale* society favour this participation.

5.1 Social organisation as a condition for community participation

Mahafale society is of the customary type where a social organisation has always been maintained. It is clear to see that the existing socio-political structures and culture have an influence on community participation.

5.1.1 Social structure

Thus, during the reign of the kings in *Mahafale* history, there was a form of participation in social and political life according to each person's place in the social structure. It was another form of participation by the *Fokonolona* in social and political life for the development of the community. This is known as *"rima*[99] *"* in *Mahafale.*

The form and degree to which responsibility is taken differs according to a person's place in the community. The *Mahafale* inhabitants live in a lineage-based society. It is divided into several clans. At the base is the lineage *(tarika),* then the clan *(Raza)* and finally, at the highest level, the *"foko".* Its structure is very broad, as the *"fati-drà"* (blood brotherhood) and the *"ziva"* (joking kinship) expand it.

However, the clan is the essential element. So property is clannish. At one time, the property par excellence was the herd. Each clan had its own herd and had its own ear tags for the zebus, the *"vilo".* To this day, these markings reveal the clan affiliation of the transhumants who move around the *Mahafale* plateau.

In the *Mahafale* plateau, this division does not leave much trace, but the social organisation in the *Mahafale* landscape is based on kinship, particularly the age group: a traditional hierarchical organisation. Each age group has had its respective role in the development of society.

The responsibilities of the *mpitata*, *mpisoro* or *mpitan-kazomanga* have already been described in detail in the first part of this book. His status gives him decision-making power in the social production process at community level. They are always consulted on all projects, including development and administrative strategies.

He participates as a referee in drawing up plans for the development of his village. His participation is a little special because he watches over the mystical aspect of development and guards against the total acculturation of the community. He is very reticent about modernisation. Consequently, he tries to influence the community's perception of development during the diagnosis.

[99] It's a collective workforce, a kind of mutual aid in the community where everyone participates voluntarily in undertaking an activity.

As for the *Ray Aman-dReny*, owners and managers of the family means of production, they make up the entire working population. They are responsible for everything that concerns society: community work and the *dina*. As such, they play an active role in drawing up the village development programme.

Young people are afraid to take responsibility for themselves and leave it to the *Ray Aman-dReny* to direct their destiny. Sometimes, during a patrol, the young people replace their parents, especially during the rainy season (work in the fields). They trust their parents with the vision and design of their destinies.

Furthermore, during the interview with the young people, some of them stated that the transfer of management belongs to the elders and not to the young people, who lack experience. They therefore think that the development of their village is a matter for their parents.

As far as gender participation in the conservation of forest resources is concerned, it has been noted that the degree of participation is different: female participation is more or less low. The general status of women is severely hampered by the pre-eminence of men. The *"mpisoro"* or *"mpitan-kazomanga"* cannot be a woman.

Cooking and household chores are entrusted to the wife, and her great skill in this area is readily acknowledged.

In Madagascar, custom has the force of law and can therefore be used to supplement it in cases where the law is insufficient or silent. In *Mahafale*, custom greatly favours the exclusion of women, not only because of their physical weakness, but also because of the dirtiness resulting from their menstruation[100], which has become their major failing throughout *Mahafale* society.

Since 1975, custom has been reinforced by the institution of the dina or collective agreement, which remains unwritten given the illiteracy of most people in the region.

The value of custom is then considered equal to the law[101] in Madagascar. This is the case of article 12 of Ordinance n°62-089 of 1st October 1962 relating to marriage, which states that: *"The prohibition of marriage between cousins, or between any other persons bound either by legitimate, natural or adoptive kinship, or by present or past ties of alliance, obeys customary rules"*[102] .

In the countryside, Malagasy positive law gives pride of place to customary law. It is not because people have a perfect command of the texts based on custom that they will respect their traditions, but it is because they believe in the traditions and habits of their ancestors. Customary rules in the south place women at a great disadvantage from birth.

Generally speaking, then, the place of women in *Mahafale* society seems very precarious. They do not attend public meetings, or have to remain in the background at open-air gatherings. In everyday life, women play a predominant role in practical activities: household chores, work in the fields and caring for livestock. Paradoxically, women, who are a major workforce, have very little access to the benefits of everyday life (education, credit, training, etc.).

---

[100] Periodic blood flow in women up to the menopause at age 49

[101] Expression of the general will of grassroots communities or Fokonolona

[102] Mahafale custom authorises and encourages marriage between cousins who are not first cousins

In particular, women are at the centre of all water-related activities (cooking, washing, vegetable gardening, etc.). What's more, in this society where the zebu civilisation was born out of a warrior production system, women do not decide the destiny of the herd.

Finally, during the group interview with the local population, one woman pointed out that everything to do with the administration and activities to develop the village are an exclusively male affair, as decision-making power rests with the men. As a result, women in the south do not feel that they are also actors in development. Yet community development and the conservation of forest ecosystems begin at home, and it is women who manage the household.

The structure of *Mahafale* society does not allow women to participate in community development or even in the conservation of forest ecosystems.

5.1.2 The political structure

Some people still think that everyone is free to do what they want in a democracy. However, they are not free to do as they please, because *"everyone's freedom ends where others' begin"*. This reveals the existence of implicit rules that govern relations between the individuals that make up society and possibly the community. These rules must reach and preserve the common good. Democracy is the ideal for this.

In Madagascar, the State and decentralised local authorities are also governed by this democracy.

Our research theme concerns the participation of local communities in the conservation of forest ecosystems. We want to find out whether Malagasy-style democracy today has an impact on village participation.

Participatory democracy, as its name suggests, calls for greater participation by citizens, indeed by the entire population, because everyone is put on an equal footing. In this sense, citizen participation reflects a power struggle. The existence of numerous political parties reinforces this struggle in our country. These parties want to take root at grassroots community level, and this desire is changing relations between citizens in the countryside.

For them, politics is simply an art of accumulating money. Politics is seen as an income-generating activity, so it's up to everyone to actively participate in order to obtain funding. The beneficiaries of the funding control and direct the community in their own way. The community suffers from peasant-style party politics, and as a result the people no longer trust the politicians. By way of illustration, a participant in the interview with the "active adults" group said that farmers can be fooled once, sometimes twice but never three times, because of their prudence.

Furthermore, rivalries resulting from the political divide prevent cohesion in community work, given the people's distrust of the leaders. This divide has repercussions on community participation in village development, hence the low level of village participation when forest resource management transfers are set up.

However, the political parties bear a large part of the responsibility for this situation of failure. They must know and play their roles as the main political players with essential and decisive functions for the population, for those in power and for the State in the process of development and democratic transformation.

They play a crucial role in transforming the political will in the country. They provide the State with political personnel through democratic elections and appointments. They must encourage the active participation of citizens in political life, train citizens capable of taking on political functions and influence political developments within state institutions.

This task is a challenge. Parties must reflect on their discredit, question themselves, learn positive lessons from party history, change their behaviour, observe the principles governing the democratic order and be able to structure themselves in the face of the multiplicity o f opinions on problems and on the possibilities for solving them. These obligations should be focused and articulated by a few relatively large parties and require an active commitment and synergy of effort from all actors: the state, political parties and civil society (including the grassroots community).

In short, democracy a priori favours citizen participation. However, we need to take a step back to see that in *Maroseragna*'s time, the solidarity of the *Fokonolona* already encouraged everyone to participate, and that society was hierarchical, unlike the democratic system advocated today.

Generally speaking, the *Mahafale* company has a traditional organisation with the following structure:

- representatives of the former royal class: these are mainly descendants of the "*Maroseragna*". The "*mpitan-kazomanga*" or "*mpisoro*" are the chiefs of the lineages who have the privilege of being able to enter into direct contact with the ancestors of the lineage and the gods of the clan. Responsible for ritual ceremonies, during which they are often assisted by an '*ombiasa*', they enable the lineage ancestors and the living to communicate directly. Often of advanced years, they are highly respected by the population.

The "*mpitan-kazomanga*" group also includes relatively independent characters:

  - the "*fahatelo*" has a social status of immediate importance after the "*mpitan-kazomanga*" and serves as the latter's assistant during ritual ceremonies;
  - the "*Tale*", chiefs of certain lineages in the *Ampanihy* region who have not yet been enthroned as "*mpitan-kazomanga*". This delay is due to the current ceremonial simplifications;
  - the "*Razana*", a kind of "*mpisoro*" who has the additional privilege of being possessed by the spirit of the ancestors;
  - *mpikabary*" or "*mpizaka*" play an essential role in the day-to-day regulation and resolution of conflicts within the traditional community. Chosen from among the elders for their knowledge of tradition, their eloquence and powers of persuasion, their negotiating and conciliatory skills, they often chair meetings.

- *mpagnarivo*" are wealthy people, from powerful families or who have made their fortune by moving away, "*ankarama*", and whose power is measured above all by the possession of a large herd. They form a powerful lobby within the community and speak loud and clear in the *"kabary"*, where their opinions are highly valued;
- *ombiasa*" comprising :

- *ombiasa*" possessed by spirits, called *"jiny"* if they are possessed by a spirit called "*kokolampo*";
- *ombiasa tromba*" if they are possessed by the spirit of a great ancestor or an important historical figure;
- and "*ombiasa doany*" if they are possessed by the spirit of a dead person buried in the lineage vault or "*doany*".

- the "*mpanandro*", a kind of astrologer, who are consulted for all the important activities of life to find out the best dates to carry them out and the worst dates to avoid at all costs;
- the "*mpisikidy*", a kind of diviner who practises divination using seeds;
- and lastly, the "*mpamorike*", a type of evil sorcerer, who are consulted for the purpose of harming others (poisoning, bewitchment, etc.).

So a person's position in the hierarchy effectively determines the form of participation they can take.

5.1.2.1 The village area

The social organisation of the *Mahafale* community is well adapted to the management of the villages on the plateau. Each village has a territory called a *"faritany"*, the boundaries of which are well known to the communities. These boundaries are marked by trees, grass, stones or other easily identifiable physical objects. They are to be distinguished from the *"vorovoro"*, which are clear indications that play the role of traditional boundaries, often temporary, used to protect reserved areas.

The *"faritany"* is experienced by the villagers not as a neutral geographical support, but as a space with which they maintain a mystical link at the same time as an agro-economic one. It is a mystical space where the sacred is strongly present, for it is in this sociological space that the dwellings of the deceased *'aritse'*, the tombs, the sacrificial poles *'hazomanga'*, those for circumcisions, the places of worship, but also the places for meetings and other manifestations of community sociality, as well as the forests used to collect amulets known in *Mahafale* as *'volohazo'* and the forests belonging to a deity known as *Tambahoake* are found.

The *"faritany"* is also an economic area that corresponds to developed or developable land, but also uncultivated land of little agricultural value, and includes grazing areas and farmland. The grazing areas are only divided by the boundaries of the village, or *"faritany"*.

Land appropriation only concerns agricultural land and is carried out by lineage. Possession often only means right of use at the level of the lineage fragment or the small family. Redistribution between brothers, heads of small families, may occur every two or three generations.

5.1.3 The culture

Having studied the political structure and the social hierarchy, and by demonstrating that democracy is the structure that favours citizen participation, we can confirm that the behaviour of a development player must correspond to the existing social structure if everyone is to be convinced to participate effectively. The values, beliefs, ideals and ideologies that underpin people's behaviour therefore influence their participation. The culture of these villagers must be taken into account because it exists to systematise the satisfaction of people's social needs, and the means of achieving this satisfaction are the various institutions, major and subsidiary, that make up the culture. For the conservation of renewable natural resources in the *Mahafale* zone, the question is whether the political culture led by the donors and the Malagasy administration in the *Mahafale* landscape is homogeneous and appropriate to this democracy in order to encourage village participation.

The political system malfunctions and is vulnerable if there is no congruence between the political culture[103] and the political structure.

Each type of culture corresponds to a type of social structure[104] . They are in harmony. For example, a parochial political culture corresponds to a traditional, decentralised structure; a culture of subjection is suited to an authoritarian, centralised structure; and finally, a culture of participation is suited to a democratic structure.

The traditional *fokonolona* institution is beginning to regain its place. This institution is moving closer to the democratic structure. However, the reality that we saw on the ground runs counter to the culture of participation, rather it is the parochial culture that seems to exist in the *Mahafale* plateau. In reality, a political culture does not suppress the previous one, it adds to it. All political cultures are therefore mixed. It includes, albeit in unequal proportions, elements of parochialism, subjection and participation.

It is then necessary to see how participation in the conservation of natural resources actually plays out at the level of the population in the study area.

For the *Mahafale*, an agro-breeding population, the forest is traditionally located at the boundary between the world of humans and that of certain divine entities, as the spirits of the ancestors, intermediaries between God and Man, live there. They also believe that the forest is the property of a deity called *Tambahoake*. The exploitation of the forest environment, which must remain minimal and punctual, is subject to certain rites and can only be carried out through the *Tompon-tany*, landowners and mediators between the divinities and mankind. Although sacred, the forest has long played only a marginal role in the *Mahafale* economic system, which is based on extensive livestock farming. So, traditionally, the *Mahafale* did not destroy the forest and protected it for essentially religious reasons.

It is true that the liberalisation of trade in land intensification and the maize trade in the south-west to meet strong demand from the foreign market, particularly Mauritius, have led to deforestation. However, in the *Mahafale* plateau, it is the *Antandroy* migrants living in the rural commune of *Beantake*, in the *Betioky* district, and the *Antanosy* from the same district, who have exploited this market on a massive scale. "*In the past, the forest was dense, with no clearings. People didn't farm much; a small piece of land was enough. Nobody sold zebu cattle, just a few goats here and there to buy rice. When people started growing maize in 1975, they*

[103] It is a level and character of citizens' political knowledge; a content and quality of values, traditions and social norms that govern political relations. Political culture has a clear class orientation. In bourgeois society, for example, the political cultures of the exploiting class and the politically exploited class coexist in struggle.
[104] These are stable and orderly relationships between the elements, the constructive parts of a society or the different areas that make it up (sciences, cultures, etc.).

started clearing parts of the forest. There are lots of lemurs, turtles and guinea fowl in the forest. [Interview with a local resident in *Bekinagna*].

5.1.4 The human capacity

Society is made up of individuals, families or households. To support these households, the population has to work hard. In feudal times, as in the *Merina* kingdom, work in the *Mahafale* kingdom was organised by family; mutual aid was the basis of this work organisation known as "*rima*". Every member of the community had to take part in community work.

At present, our study concerns the conservation of renewable resources through the transfer of management of forest ecosystems in order to control deforestation but also to reduce anthropic pressure on *Tsimanampesotse* Park. The work organisation in this management transfer is different from that of the *rima*.

The form of participation in the management of forest resources through the transfer of management is different from that of community work. The transfer of natural resource management requires the participation of the whole population.

Furthermore, Madagascar's capacity shortage index was 0.353 in 1993, according to the 1996 Human Development Report. The female participation indicator was 0.481. These figures show that a large proportion of the population suffers from a lack of capacity and that there is very little participation by women in the management of natural resources.

The data below reveals the intellectual poverty that prevents people from participating. People with intellectual limitations, given their level of education, cannot participate actively in the political, social or economic life of their locality.

Figure 10: Level of education of respondents

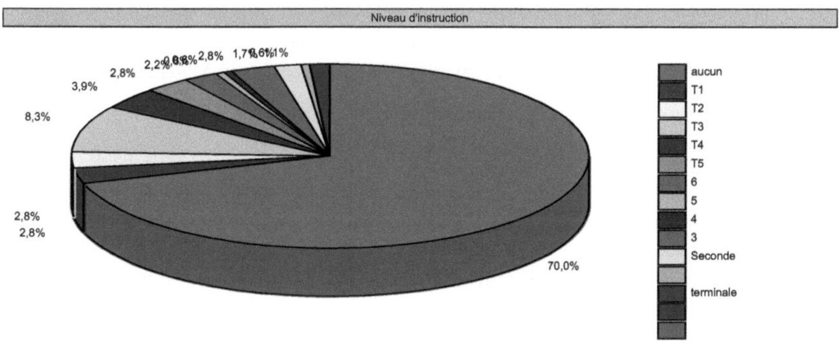

Source: Our own survey, 2014

This figure shows the educational level of the respondents in our research area. We can see from this data the intellectual level of the inhabitants of the *Mahafale* plateau. Thus, 70% of them have no education at all, i.e. they have never been to school. While 8.3% of those surveyed have passed the 9th grade (T3 in Malagasy). 3.9% of respondents had completed grade 8 or T4. Then, 2.8%

of respondents completed the 12th grade (or T1). The same applied to those who had completed 11th grade (or T2), 3rd grade and 7th grade (or T5). 2.2% of respondents had completed 6th grade. And 1.7% had completed 2nd year. 1.1% of respondents reached the final year of secondary school. 1.1% of respondents went on to the 5th year. Lastly, 0.6% went on to 4th grade and 0.6% of respondents went on to 1st grade.

These figures show that the population in our area has a very low level of education. For those who have attended school, most of them only reached primary school below the 7th grade.

However, the donor, through its intermediary partner, is trying to train the individual to take greater responsibility. People are involved in devising strategies and working conditions. Personal aspirations are reconciled with the demands of production so that all participants adhere to the objectives, control themselves and combine their efforts in mutual understanding.

One of the current objectives of the donor is the co-management of ecosystems between MNP and the communities. Having undertaken the extension of the park, the donor and MNP agreed that a structure called CLP[105] should be created to carry out joint patrols with MNP officers. This collaboration symbolises the co-management of management transfers and the park. Park officers can no longer control this protected area on their own. Joint patrols are used as an effective method of controlling the park. Consequently, the local population, as citizens, are expected to participate directly in the management of the transferred resources and in the protection of the park, while at the same time having other priority occupations. As a result, citizen participation is blocked either by their limited intellectual capacity or their physical incapacity.

5.2 Climatic conditions as a condition of participation in conservation

"Miola ty andro henane" literally means that today's climate is uncertain. This is what the communities of the *Mahafale* plateau are saying. They have sensed this change through global warming. Indeed, every year the drought persists. It systematically affects half a million people in the south. It is also leading to a proliferation of respiratory diseases, which are caused by the use of charcoal stoves as a source of energy by almost all households in Madagascar.

Communities need to come up with a plan and ways of adapting to global warming.

According to the IPCC[106] , adaptation is the adjustment of natural or human systems in response to current or expected climatic stimuli or their impacts, which moderates the damage or exploits the beneficial opportunities it generates. An adaptation plan is therefore the set of actions required to enable an ecological or socio-economic system to anticipate and adapt to the direct and indirect, short-, medium- and long-term impacts of the climate, or to exploit its potential benefits.

Its aim is to strengthen the resistance and resilience of these systems in the face of these changes.

5.2.1 The effects of climate change

There are close links between insecurity and the threat posed by natural disasters to the well-being of individuals, households and communities on the one hand, and poverty on the other.

[105] CLP is a local park committee. This committee is recruited from the CoBa office member in their locality to help park officers patrol the park through the transferred forests.
[106] Intergovernmental Panel on Climate Change (2007)

There is a direct relationship between poverty and the severity of the impact of natural disasters, such as drought, floods, cyclones or migratory locusts. This relationship manifests itself at two levels. The first level relates to behaviours that tend to exaggerate the extent of the damage when a disaster occurs. For example, the deforestation caused by agricultural clearing and the excessive felling of trees in anticipation of cyclones. In urban areas, the uncontrolled and sometimes illegal occupation of areas unsuitable for habitation (close to rubbish dumps, low-lying land used as rainwater outlets) by poor people exposes them to increased risks of disaster (epidemics, flooding).

The second level of relationship is that the degree of poverty and vulnerability are intimately linked. In the case of climate change, vulnerability is defined as the degree to which a system is likely to suffer or be adversely affected by the adverse effects of climate change, including climate variability and extreme events.

Climatic hazards are increasingly unpredictable. New farmers in practice are easily discouraged when they have faced these hazards in their first year of agro-ecology. They tend to blame the new practice for the consequences of bad weather seasons, not to mention unfavourable beliefs.

The poorer people are, the more difficult it is for them to protect themselves against the harmful effects of weather, climate and health disasters, and to recover from the losses they suffer. Ultimately, poverty and natural disasters are mutually reinforcing: poverty increases vulnerability to disasters, while the frequency of disasters exacerbates poverty by destroying the productive capital available to the poor. This aggravation is caused by direct losses of productive assets held by the poor, but also by indirect losses resulting from the fact that the frequency of disasters discourages investment (for fear of suffering further disasters). Poor people who make their living in the informal sector are particularly vulnerable because disasters affect not only their own assets, but also distribution channels and sources of supply. Given that the informal sector is a major provider of employment and livelihoods in Madagascar, its extreme vulnerability can have major adverse consequences for the poor, particularly in urban areas[107].

Figure 11: Comparison of production 10 years ago and today

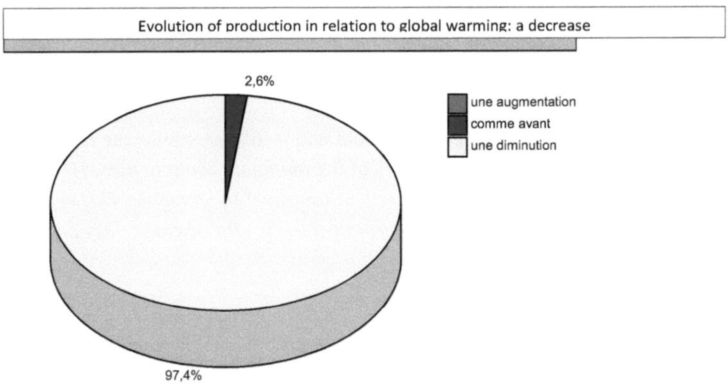

Source: Our own survey, 2014

[107] Gestion durable des ressources naturelles et réduction de la pauvreté-Préparation du PE III 06 pages.

Figure 12: Comparison of production five years ago and today

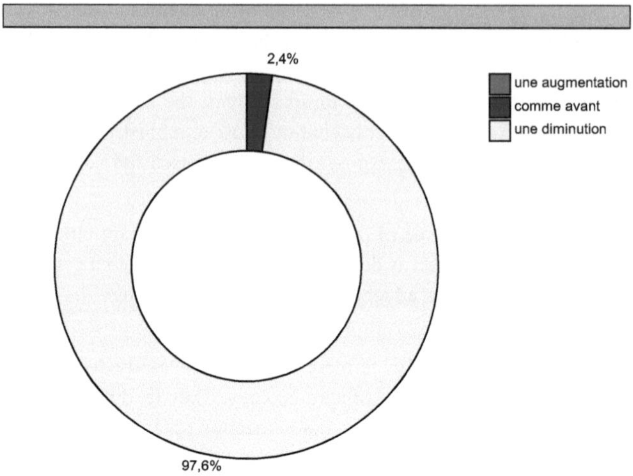

Source: Our own survey, 2014

The two figures above show the constant decline (year on year) in the productivity and products of communities in the *Mahafale* landscape due to the disruption of rainy periods: only three months of rain and nine months of drought. When asked about production compared with ten years earlier, not a single respondent admitted to an increase in production. On the other hand, 97.4% said their production had decreased. 2.6% saw no change in their production compared with ten years ago.

Then, when asked about their production compared with five years earlier, 97.6% of respondents confirmed that their production had decreased, and 2.4% found no change.

These figures show that there has been an effective permanent fall in products, even though the difference over 5 years is 0.2%. The latter is equivalent to an increase in daily minimum temperatures, the shortening of the rainy season[108] and the violence of the winds brought by tropical cyclones over the last 20 years.

In other words, global warming is beginning to take hold and is also preventing the regulation of rainfall, which is disrupting the cropping calendars of the *Mahafale* communities. "*We used to plant even during the dry season. This practice is called kantray*[109] *in Mahafale. This enabled us to alleviate the food shortage during the rainy period between rainy seasons. This practice*

[108] Rainy seasons have become shorter as the maximum number of consecutive dry days in the year has increased by around one day per year over these 45 years. The total annual amount of rain has also fallen over this period. With rainfall falling for shorter and shorter periods, the intensity of rainfall has increased, particularly in the northern half of the region (In Témoignages de Madagascar: "changement climatique et modes de vie ruraux", WWF 2010).

[109] The fields are ploughed and cultivated towards the end of the dry season. If the arrival of the rain is delayed, the crops dry out.

is no longer possible, as it would simply be a waste of seeds. It used to rain heavily in January. Now there's not a drop of rain this month. [Interview with the President of CoBa *Mizakamasy*].

5.2.2 Land factors

A few land-related disputes have been recorded, including disputes between herders and farmers, and theft of cattle between families. But these disputes are generally settled out of court either by the mayor or by the *Mpitan-kazomanga*. However, as *filongoa or fihavanana* is more dominant than the law, it has to be said that existing disagreements tend to be referred to the *Mpitan-kazomanga* before being brought before the legal authorities.

It is very rare to encounter major land disputes, but when they do arise, everything is settled out of court, even in the presence of the *dina*: when an arrangement is reached, the *dina* are ignored. Each region has its own *dina*, and the texts differ according to the priorities of those who established them. Some *Fokontany* also have dina, and their application depends on the higher level, such as the town hall or the courts.

In our study area, there is as yet no land tenure office. According to those in charge of the National Land Programme, this project is in the pipeline, but local players have never mentioned it. As for the people in charge of the land services, no fixed programme has been organised beforehand, and visits to the land depend mainly on the number of applications in progress, which explains why local people are unable to really understand the details of the procedures for demarcating land.

According to the local authorities, the high cost of the process and the illiteracy of the farmers prevent them from securing their land. The South-West Region is going to ask the government to set up a provisional land office at regional level. The aim is for each commune to have a local land use plan. But producing a plan is prohibitively expensive, costing around 4,000,000 Ar. At this stage, the rural communes are blocked.

As far as access to land is concerned, all that is required is to inform the local authorities, i.e. the *Fokontany* chief, of a wish to settle on the land. Proposals are made to the applicant after a meeting and with the agreement of the *Fokontany* committees, and the migrants or investors choose. There is no time limit, but the investor is obliged to contribute to local development. The same applies to foreign investors.

The legal knowledge of the inhabitants of our study area is very moderate, and customary and practical rights are still very much respected.

Customary law is strongly applied compared to positive law. In fact, positive law exists while people inexorably rely on customary authorities before thinking about legal authorities.

It should be remembered that the Malagasy give an important place to ancestral beliefs: customs and practices as well as customary rights are always taken into account. Sometimes, the laws in force are forgotten or even disappeared. Lack of information is one of the factors blocking understanding of the laws.

However, ignorance of the texts and decrees is not considered to be a serious matter by the inhabitants, either for their homes or for their wealth. In practice, the inhabitants consider that all of the land belongs to them, whereas *"State legislation stipulates that the State is presumed to be the owner of any land that is not registered, cadastralized or appropriated by virtue of regular concession titles"*. [RAZAFIARIJAONA J., 2000], and small land conflicts, generally concerning the demolition of crops by livestock, remain temporary and unimportant. The *filongoa* is there to solve the problems.

Furthermore, the "titling" of the one and only property in a rural commune such as *Beahitse,* for example, originates in a family dispute. Even the mayor of this village does not bother to demarcate or register his own home; a simple fence is enough.

In this sense, these local authorities are preparing to mention land tenure security as an objective to be achieved in their communal development programme, while the person mentioned as being primarily responsible for raising awareness of the usefulness of land titles feels absolutely unconcerned. Were these documents drawn up in haste, or were they undertaken carelessly?

Similarly, the park still has no land title, even after its extension. MNP officials are not yet planning to secure the protected areas.

The use of the land by the inhabitants and their livestock is a very important and recurring fact, symbolising a kind of demarcation so that none of the territorial groups who move in can say who owns the land. Historically, the *Mahafale* seem to have been the first occupants, but the fact that they belong to a social life that is strongly linked to the various customary aspects, and also the current mass migration of other Madagascans, prove that securing land tenure is essential.

Apart from the Chinese who are going to extract the *Sakoa* coals and those who have explored the oil deposit in the *Betioky* region, there is a group of Chinese who are entering the north of the park, on the *Ereteke* side, without having notified either the *Betioky* district authorities, the Commune of *Ampasindava* or the management committee of CoBa, which manages the area. It must be reiterated that entry into the management transfer zone, such as the park zone, is strictly forbidden for everyone. A pass is required.

The Malagasy government is currently looking for foreign investors. But when the latter start to take an interest and explore the subsoil, will we have to wait for a large sum of money to be offered by these foreigners?

The *Mahafale* plateau covers an area of 1,744,200 ha. Twenty-seven per cent (27%) of the plateau is protected: national park and management transfers. Arable land accounts for no more than 15%[110] of the plateau's total surface area. Compared with the communities to the east of the park, known as *Ankara*, the coastal dwellers suffer from a reduction in space imposed by the introduction of the management transfer and a marine protected area. They do not live freely on the land of their ancestors.

Since 2009, there has been a massive immigration of young people due to the reduction in their exploitable areas and the drought. They have chosen two different places to take refuge and rebuild their lives: towards land where there is sapphire or towards *Morondava* to farm. For those who cannot immigrate, they are forced to cultivate clandestinely within the park that once sheltered both the tombs of their ancestors and the communal pasture. "*Behind our village lies the sacred Rehindy forest, where the spirits live and our noble ancestors rest. A third of the population didn't want to leave because for them leaving the village meant abandoning their ancestors, their roots and their cultural values to live as strangers and refugees in other people's homes.*" [Interview *Mpitan-kazomanga Bekinagna*].

---

[110] In Testimonies from Madagascar: "Climate change and rural lifestyles", WWF 2010

This situation calls into question the implementation of the Gelose law in the *Mahafale* plateau. The *tompontany* (landowners) feel marginalised and the conservation of forest resources does not solve the basic needs of the local population. For the three thousand inhabitants of the rural commune of *Beahitse*, for example, there is only one well in the village, and on the coast, the water considered drinkable is very salty. The lack of water is leading to the degradation of biodiversity and to uncitizen practices. Indeed, several main roads in the interior of *Ankara* are suddenly blocked by fences of cultivated fields, as it is easy to pipe water along the road to the newly planted fields. This practice reflects human pressure on natural resources [VERIZA R.F., 2005].

So it makes sense to ask why transhumance has been known about for a long time, and why the shortage of water, drinking or otherwise, in the south is nothing new when no one is managing to meet its needs? Are the local people responsible for this too?

Chapter Six: Alternative activities for the conservation of natural resources

According to Burney, deforestation in Madagascar is the cause of multiple attacks (climate change, repeated burning of plant cover, biological invasions) acting in synergy on the most sensitive areas. However, faced with the anthropic pressure on the *Tsimanampesotse* park and the forests transferred to the communities, the COGESFOR project funded by WWF has implemented a strategy that involves working with local communities to develop and implement alternative solutions to these practices.

The main aim of the project is to preserve the environment while improving beneficiaries' incomes by improving farming practices and livestock management (particularly housing and feed). In this way, the project aims to improve agricultural production for farmers in the project areas throughthe emergence of a consensus on how to manage pastoral and natural resources in the local areas, in particular through better integration of agriculture, livestock and the environment. he interactions between these three elements (cultural, nomadic and sedentary pastoralpractices, which havea significant impact on the environment) require action that takesa more global view of production systems,from an agronomic, geographical and social perspective.

6.1Capacity building in new cropping techniques (SCV)

Drought is rife in the south and south-west of Madagascar. Agriculture and livestock farming are the sectors most affected. The *Mahafale* plateau has a very low population density. It is made up of poor agro-pastoralists who raise cattle and move from one place to another. According to the AVSF report on agro-ecology in Madagascar, over 55% of the plateau's population do not meet their basic food needs.

Many farmers have not been able to plant every year, as it is difficult to find seeds for the season, which has had average rainfall. Despite the small rains of the previous year, in most cases the crops planted by the villagers will not survive. Agricultural problems such as clearing and slash-and-burn[111] or *hatsake* and overgrazing of savannahs and dry forests, the reduction in annual rainfall with strong inter-annual irregularities and the savannisation of vegetation and aridification of the environment are on the increase. As a result, the heads of environmental NGOs fear that the park will come under heavy human pressure.

---

[111] For Madagascar in general and the *Mahafale* plateau in particular, the slash-and-burn system was probably introduced by the first people to arrive on the island in the seventh century. It was therefore introduced either by Indonesians and/or Polynesians from South-East Asia, Africans from East Africa, or Arabs or Indians from Central Asia or the Middle East. Cultivation follows the same procedure throughout the world. The land is cleared, i.e. the trees are felled and then burnt, but the stumps are left in place. The cleared plots are cultivated for one, two or three years, rarely more. After that, they are abandoned to wooded wasteland for one or more decades before being cleared and cultivated again. The name changes for each ecoregion or territorial group: *Tavy* for the *Betsimisaraka* in the east, *Teteke* for the *Antandroy*, *Hatsake* for the *Masikoro* and *Tanalana* in the west...

When the population grew in numbers and rotation and crop rotation were no longer respected, slash-and-burn farming became a bad idea, because it destroyed the forests that regulated atmospheric flows and the climate. Clearing by farmers removed the water contained in the biomass and the moisture locked in the soil. This led to an upheaval in rainfall and a warming of the surrounding climate.

The project commissioned AVSF to popularise agro-ecology and conservation agriculture based on SCV (no-till sowing on plant cover). According to Gliessman, agro-ecology is inspired by the functioning of natural ecosystems and traditional forms of agriculture. In these systems, productivity increases because of the efficiency of living organisms in recycling nutrients.

To disseminate conservation agriculture on *Mahafale* land, the project has set up demonstration plots of cropping systems and DMC systems and has supported pilot farmers around these demonstration plots. The logical framework set 10 demonstration plots and 30 pilot farmers as objective values. AVSF was able to support 45 pilot farmers, including 30 active farmers, on 5 ha of demonstration plots. These farmers are those who are sufficiently well off economically, informed and interested in intensifying rainfed production (labour and/or inputs).

However, in several Malagasy regions, including the *Mahafale* landscape, farmers are already practising crop associations and fallows and/or using trees to restore the fertility of their plots. As with the vast majority of peasant farming in the world, agro-ecology is therefore nothing new, even if some of the innovations currently being promoted by research and development can greatly improve these practices, or develop systems that are sometimes better suited to the constraints faced by peasant families. On the other hand, it has to be reiterated that the fundamental problem facing communities in the *Mahafale* region is drought. They use obsolete seeds, but the use of livestock manure (cattle, goat, sheep and even chicken) is still *faly* (forbidden) among the *Mahafale* because it is a waste product that should not be eaten.

The crop grown by the farmers did not survive. Similarly, the demonstration plots set up at the demonstration site will not survive if the few rains received at the start of the project are not followed up, even if the project uses improved seeds.

Farmers have asked the project to supply them with improved seeds so that they can withstand climate change, as the often observed shift in their practices towards unsustainable cropping systems is often linked to uncontrolled factors (little quality land available, land insecurity, weed pressure, lack of plant biomass and organic manure, pest pressure).

The project has not thought about the strategy for disseminating the new techniques so that they are not limited to the members of the CoBa (a few pilot farmers) but are also adopted by other farmers who have difficulties with them, in particular those who have had their access to resources restricted by the introduction of TGRN contracts. The project's objective is clear: to demonstrate the new techniques to the communities. "*The project is unable to meet people's needs at this level, because we're here to demonstrate techniques, not a seed distribution project.* [Interview with WWF project manager].

The second component of the project, which justifies the implementation of conservation agriculture, is the development of environmentally-friendly value chains (timber, aromatic and medicinal plants or other promising sectors). Our question in relation to the implementation of this component is: is this project being carried out for the park's local communities or to satisfy the donors?

When we spoke to the head of AVSF, based in *Betioky*, he told us that there was no budget for implementing the project, as WWF is only contributing 9,000 euros a year to this conservation agriculture extension project. This is an inadequate contribution given the number of activities to be carried out and the size of the target area.

The visit and discussions with the pilot farmers in our field area enabled us to understand that some of them were able to continue the work timidly despite the lack of support from the project: the latter decided to break the contract with AVSF after the evaluation demonstrated the negative results of the agro-ecology promoter. A kind of internal organisation was set up among the farmers. This solution has enabled them to save seeds and exchange seeds among themselves. At present, insect and bird attacks are the main problem for pilot farmers, and unless a biological solution can be found, they are not allowed to use chemical control methods.

We have also noticed that the low level of schooling represents a major constraint to the dissemination of the method. In order to better understand the various specific features that make the method so successful, a minimum of schooling is very useful. In addition, it was found that the pilot farmers who attended school the most were the easiest to supervise and adopted the recommended techniques the most.

WWF should then find a solution to resolve the above problems.

Why is AVSF so negative?

Apart from what we have said, the project had no system for monitoring the dissemination of the techniques extended by the pilot farmers and was unable to set up a tool for steering the component's action. It left the extensionist to implement the component on his own. No specifications have been drawn up for these pilot farmers. This is obvious because the project has not been able to set up tools to monitor AVSF's activities. As a result, there was no file available for capitalising on the intervention and its achievements. As a result, the project was unable to assess the AVSF's activities in order to direct it towards dissemination actions on systems that were better adapted to the environment and communities. In short, the complete and reliable database of these 30 pilot farmers did not exist. "*As a result, we were unable to carry out technical and economic analyses and assessments of our interventions*", admitted the COGESFOR project manager.

What's more, as the project only lasts a few years, four to be precise, the continuity of activities to ensure the sustainability of its actions leading to a change in the behaviour of communities to adopt agro-ecology is not at all well organised. The project's vision of this aspect was too limited to public services. In reality, as a result of the government's policy of disengagement, these public technical services no longer exist in the project's areas of intervention. Even if the private infrastructure that can supply our field area with seed is very inadequate or almost non-existent, the project will have had to exploit its knowledge of the local area by canvassing civil society organisations that can relay its extension activities, such as the MdP, before leaving the pilot farmers to fly off on their own. In fact, it is a local structure that already has technicians and farmer trainers known as 'technical delegates' who will have had to identify and train. The advantage of such a structure is that it is local and will be able to sustain the actions.

As the communities have requested the supply of improved, drought-resistant seeds, the project should negotiate with economic operators or wealthy people who could open an input and seed supply centre in the *Mahafale* landscape if the project intends to popularise these conservation farming techniques.

6.2 Literacy training for grassroots community members

As we said earlier, most of the grassroots communities in the *Mahafale* landscape are illiterate. In the PCD of the urban commune of *Ampanihy*, 85% of the population are illiterate. The statistics for the two districts of *Toliara* II and *Betioky* are similar to those for *Ampanihy*. Managing a renewable natural resource such as the forest requires a minimum level of education, if only to draw up an activity report. The level of education is one of the requirements of environmental operators, in particular the GIZ, for good governance involving good management of forest resources. Grassroots communities therefore have the right to be literate in order to improve the management of natural resources. If a community is to commit itself to the sustainable management of the natural forest, it must have a guarantee that it can continue to benefit from its management and exploitation rights as long as it complies with the conditions of the transfer contract.

Figure 13: Level of education of CoBa members surveyed

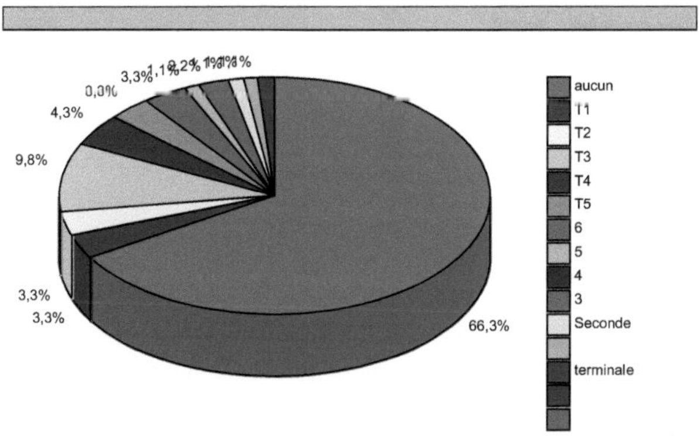

Source: Our own survey, 2014

This graph shows the level of education of the members of the CoBa surveyed. Thus, 66.3% of those surveyed have no level of education. They are illiterate. 9.8% of respondents had completed 9th grade. 4.3% of CoBa members surveyed had completed the 8th grade. 3.3% of CoBa members surveyed had dropped out of school in the 11th, 10th, 7th and 6th grades. 2.2% of respondents had reached the 4th grade. 1.1% of CoBa members who took part in the survey had attended collège and lycée, i.e. the 5th, 2nd,^{nde} , 1st and final year of secondary school (^{ère}).

To sum up, of the CoBa members surveyed, 66.3% are illiterate, 24% have primary education and 10% have secondary education. They are made up of illiterates because even though 24% have passed primary school, most of them have forgotten how to read a sentence and how to write. There are no university-level members in CoBa.

The WWF project realised that the communities were not working in the desired direction, even though they were generally aware of the conservation issue themselves or through the interventions of conservation organisations. As a result, the project decided to undertake a literacy campaign among CoBa office-holders. This literacy project came to a halt because the office members who wanted to become literate had to pay school fees, even though they were poor and their priority was to buy food but not to become literate. They are not motivated to pay money to learn to read and write. "*The project needs to understand that we guard WWF forests and we should learn to read and write for free. The project should improve our collaboration*" [Focus group with members of the *Ampitanake* CoBa].

Figure 14: CoBa membership by level of education

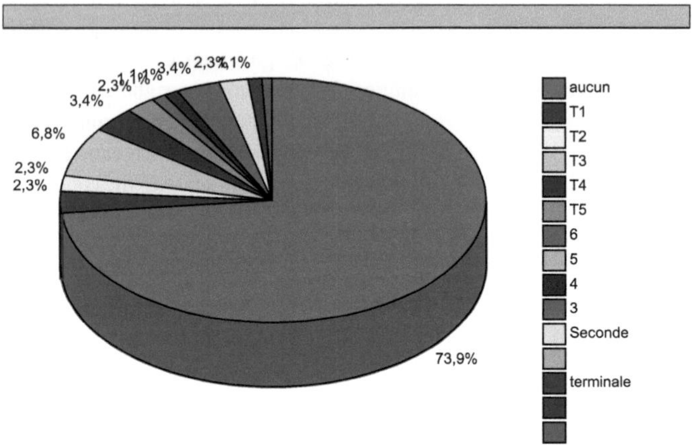

Source: Our own survey, 2014

This graph shows that the level of education influences membership of the CoBa in the *Mahafale* plateau. Thus, 73.9% of illiterates, 17.1% of primary school pupils and 9% of secondary school pupils are not motivated to join a grassroots community to manage forest resources. It should be remembered that in the PCD of the communes of the *Mahafale* plateau, more than 80% of the population are illiterate.

But if we compare the level of motivation of educated and illiterate people, the latter are more motivated to join CoBa than the literate because they think that joining CoBa is their last resort.

Apart from illiteracy, the closed nature of tradition (habits and customs), the lack of arable land, drought and poverty have weakened the communities of the *Mahafale* landscape and they are currently at an impasse. However, in general, most illiterate people do not want to join the CoBa because of hesitation and fear. Fear of being abused or of being complicit with the rulers.

6.3 Development and protection of the environment

The concept of development is fairly common in economic literature and in the language of economists. But for some, development remains an ambiguous concept and a subject of discussion, since it has been the subject of innumerable reflections, studies, clarifications and criticisms that have given rise to numerous theoretical contributions. But defining development is a difficult problem, with contradictory answers. Is development the same as growth and/or human development and/or capacity?

The concept of development was initially established to be applied to a part of humanity, the part that was destined to grow up, the part that was a child or adolescent during the war and that had to be helped (as one helps a child) to reach maturity.

Development was originally based on economic growth. This refers to the increase in the volume of goods and services produced by an economy over a given period, and is the only way to achieve development.

Beyond its economic, social, cultural, spatial and sustainable dimensions, the word "development" in the strict sense is often interpreted as a process of transformation that accompanies growth over the long term. This is measured not only by the level of consumption, but also by the level of education and health of the population, as well as by the degree of environmental protection, i.e. the satisfaction of basic human needs in a long-term cumulative process. Economic development in the broad sense of the term (lato sensu) therefore includes other elements as well, in particular progress towards equality of opportunity, political freedom and civic liberties.

To take this a step further, Amartya Kumar SEN[112] has introduced the 'ethical' dimension into the notion of development. Every human being has the right to dignity. This is the principle of the dignity of the human person, according to which a human being must be treated as an end in itself. This principle is fundamental in the context of cooperation, as it requires respect for others, their differences and their values.

Growth also promotes development by creating wealth and raising individual living standards. Growth is the increase in GDP from one year to the next. This gain is supposed to reflect an improvement in well-being. Growth therefore sanctions development because it benefits from the rise in investment and consumption.

Compared with growth, which is purely economic and quantitative, the concept of development covers all the changes that affect both the economic sphere (production, consumption) and the social sphere (sectors of activity, professions and socio-professional categories, culture (education, lifestyle). It is with this in mind that a country's development is measured in economic terms and must take into account the human being, who is the focus of all development action. In 1990, the UNDP developed a composite indicator, the HDI, to compare the level of development of countries. This indicator groups together several criteria: GDP/capita 4 (standard of living); the level of health of the population (measured by life expectancy at birth/ 55.5 years for the Malagasy); the level of education of the population (measured by the adult literacy rate, for 2/3, and by the average number of years of study, for the remaining third).

While it is possible for growth not to go hand in hand with development (when redistribution does not take place at national level, as in certain oil-producing countries where growth is strong but society is still full of archaisms), there can be no development without growth: improving a population's living conditions cannot be achieved without significant material resources.

According to the economist Walt Whitman, the dynamic phenomenon of growth is not automatic and immediate, but takes place in stages[113] .

---

[112] Nobel Prize in Economics in 1999, human development economics

[113] For the traditional society, the stages of growth are the society is agricultural, stationary where the land is the

Broadly speaking, then, development can be defined as a coherent and harmonious political, social and economic process that generates a state of life, being and thinking that is conducive to a sustainable and desired improvement in living conditions. Consequently, it is clear that "*development is a combination of mental changes in the population that enables it to increase its overall production cumulatively and sustainably*" [François PERROUX, 1964].

6.3.1 Local development

The origins of local development lie in the ineffectiveness of macroeconomic policies and national sectoral measures in resolving day-to-day economic and social development problems at local and regional level.

The local development approach is original [VACHON B, 2001] because it makes it possible to mobilise and stimulate the dynamic elements and resources of the community with a view to sparking new projects, triggering and supporting individual and collective processes of change and development. The impetus does not come from outside, but from within. To achieve this, a series of actions should be undertaken to put the region in a position to develop.

The endogenous and exogenous levels must complement each other. Endogenous local development does not exclude aid from "above". Indeed, the first level (endogenous) mobilises the population, stimulates innovative ideas, develops projects, makes the most of available resources, and enhances the will and capacity to act, while the second level (exogenous) provides aid in the form of structuring investment, training, funding, technical support, local power, and so on. Local development thus appears to be the meeting place between what comes from below and what comes from above.

Consequently, a local development project is one that is initiated and implemented by and with the participation of its residents and their "institutions". It attempts to respond to their aspirations, and to make the most of their collective "assets".

Indeed, sustainable management of natural resources can form the basis for community and local development, provided that natural resource management is based on a participatory approach that enables local stakeholders to take charge of and take part in environmental management, and that development takes into account economic, environmental, social, political and institutional aspects.

In this context, the local community should be understood as a group of people who are territorially located [BALLET J., 2007]. In fact, it appears to be the appropriate level for resource management and is considered to be the closest user of these resources. The territorial concept is therefore a determining factor in the collective identity of the people who live there. This concept makes it possible to define and effectively implement development actions in a participatory manner.

In this thesis, our approach gives concrete form to all these aspects by analysing the sustainable natural resource management system involving the CoBa, thanks to the implementation of the biodiversity resource management transfer, with which this CoBa can benefit more from the economic valorisation of renewable natural resources. The emergence of this local dynamic is therefore a support for development.

---

only source of wealth; the level of use of technical progress remains relatively low and there is a low level of productivity of economic sectors. The prospects for change are non-existent.

6.3.1.1 Sustainable development

The concept of sustainable development was first defined in a report commissioned by the United Nations in 1987 from the World Commission on Environment and Development (WCED). Taking into account the two divergent concerns of economic development and biodiversity conservation, the Commission's main task was to consider and address both environmental issues and those relating to economic growth. This opposition led to the advent of the concept of sustainable development, which is incompatible with environmental degradation. The WCED emphasised that: *"Sustainable development meets the needs of the present without compromising the ability of future generations to meet their own needs*[114] ". This concept seeks to reconcile areas as different as the economy and ecology, while ensuring a fair distribution between generations and between North and South.

Faced with the alarming degradation of the environment, linked on the one hand to the economic growth of the countries of the North, through the increase in the extraction of natural resources to supply their industrial activities with raw materials, and on the other hand to the massive use of nature to meet the needs of poor countries, international environmental awareness began to emerge with the aim of seeking solutions together. From 1992 onwards, this awareness became widely known with the Rio Conference, which decided to combine environmental conservation and economic development. Bringing together representatives from 172 countries and various governmental and non-governmental organisations, the conference marked a growing awareness of the need to manage environmental issues on a global scale in order to ensure the sustainable development of humanity.

6.4 Conservation through development: developing the coal industry

The charcoal industry is a major challenge for the sustainable management of natural resources in the South West region. Apart from the fact that this activity is one of the main causes of deforestation in the region, the charcoal production techniques currently used are leading to the rapid disappearance of certain forest species such as the *kily* or tamarind (tamarindus indica), which also has a high cultural value in the region.

The vegetation in this part of Madagascar is characterised by dense xerophilous dry forests. The area's soil and climate conditions do not allow the forest cover to recover quickly enough once it has disappeared. For this reason, it is far preferable to preserve the existing forest cover than to think about restoration or reforestation.

6.4.1 Transformation by carbonisation

Carbonisation is carried out using traditional millstones. The process is delicate and requires constant supervision by the charcoal-maker, who must avoid accidentally setting fire to the wheel. Controlling the circulation of air in the grindstone is one of the most important conditions and must be part of the charcoal-maker's know-how.

Carbonisation is carried out in the forest: the carbonisation area is chosen so as to avoid having to travel as much as possible to harvest the wood and transport the charcoal produced. The

[114] BRUNDTLAND Commission (commission of international experts chaired by the Norwegian Prime Minister GRO HARLEM BRUNDTLAND).

grindstones are erected on the ground rather than in pits. The orientation in relation to the prevailing wind is not determined. The fire is controlled and managed from holes made in the side walls of the millstones using a wooden rod or the handle of a spade. The size of the millstones used, which are generally parallelepipedic in shape, varies greatly: the apparent volume of the millstones varies from 3 to 36 m^3, the length from 2 to 6 m, the width from 1.5 to 5 m and the height from 0.8 to 2 m.

Charcoal-making does not distinguish between categories of wood, so all species are used, except those with a low density (e.g. *Famata, Daro, Kapaipoty*).

In short, charcoal producers use traditional techniques and are obliged to fell a lot of trees to get a cart, so they do not select species for carbonisation.

6.4.1.1 Socio-economic characteristics of coal miners

Coal mining on the *Mahafale* plateau began in 1982. More than 90% of the area's coal miners are ethnic *Mahafale.* The vast majority of those involved are immigrants. They are immigrants from neighbouring Communes and are in a precarious situation, as shown by the following data:

- 54% of these charcoal burners grow cassava alone, on an average area of 1 ha (0.5 to 5 ha);
- Only 8% grow crops other than cassava;
- 34% have no farmland. With the recurring problem of insecurity, despite the pastoral vocation of the dominant ethnic group, only 12% of them still rear cattle.

On the one hand, and this was confirmed by what they said during the surveys, charcoal-cutting has recently become a necessity for survival, and does not really represent a "profession", but a stopgap measure. Their submission to the traditional authorities, even before the cantonment, is still very much alive, and the use of resources is subject to the authorisation of these traditional authorities, before the cantonment authorisation is obtained. The so-called "forbidden" forests are still protected and respected, and the existence of a structure among the charcoal burners can be seen as an asset for the sustainable management of the remaining forests.

6.4.1. 2 Production

This first stage involves cutting the wood in the forest, transporting the cut wood from the felling site, cutting the branches before kiln-firing and kiln-firing itself.

For every 100 trees felled, the farmers get 3 cartloads of charcoal, and a cartload of charcoal can yield 3 to 4 sacks, so for every 100 trees, they get a yield of 9 to 12 sacks. To this end, the charcoal-makers have realised that each year they have to go a little further than the previous year to find trees to cut down.

Faced with this situation, during our interview, the DREEF was keen to point out that the problem of charcoal has grown in recent years and that measures would have to be taken fairly quickly to control the destruction of the relics of forest that currently exist. In any case, everyone was aware that this is a very delicate issue, requiring a well thought-out approach, particularly in view of the political context currently prevailing in the country.

The table below shows the area of forest used for both local consumption and for export to *Toliara.*

Table 10: Minimum annual area by Commune

Municipality	*Ambatry*	*Beantake*	*Masiaboay*
Export *Toliara* (Ha)	117	877	123
Local consumption (Ha)	17	65	180
Total surface area (Ha)	**134**	**942**	**303**

Source : ONG PARTAGE

According to this table, three Communes are the best known charcoal producers: *Ambatry, Beantake* and *Masiaboay. Beantake* is the leading charcoal-producing commune in the *Mahafale* plateau. It is therefore the commune that destroys the most trees per hectare. Every year, charcoal-makers in the Commune destroy 942 hectares of trees to make charcoal. Charcoal-makers clear 877 hectares of trees to supply the town of *Toliara*, compared with 65 hectares for local consumption. In short, charcoal from the Commune of *Beantake* is used to supply the town of *Toliara.*

Similarly, in the Commune of *Ambatry*, charcoal burners destroy 17 hectares for local consumption, compared with 117 hectares to supply the town of *Toliara.* In a single year, the Commune of *Ambatry* destroys an area of 134 hectares.

In the Commune of *Masiaboay,* 303 hectares of forest are destroyed to make charcoal. The charcoal-makers are more concerned with local consumption than with supplying *Toliara.* 180 hectares are used for local consumption, compared with 123 hectares for supplying the town of *Toliara.*

Photo 5: Sale at the Betioky daily market

6.4.2 **Controlling uncontrolled coal mining.**

The city of *Toliara* is supplied with firewood and charcoal mainly from natural forests. Above all, the need for charcoal in urban centres continues to grow. The main routes for supplying *Toliara* with domestic fuel were the *Miary* route, the RN 9 route and the RN 7 route, the latter two supplying 90% of the city's needs[115] .

A year later, an extension of *Toliara*'s wood energy supply zones was noted towards the RN 10 axis of the *Mahafale* plateau. These pressures were beginning to spread into the new extension of the *Tsimanampesotse* Park in the *Mahafale* Plateau landscape. 3 communes are affected: *Beantake, Masiaboay* and *Ambatry*. They are all in the District of *Betioky*. This landscape, rich in endemic natural resources, is very fragile and difficult to restore due to the harsh climatic, physical and edaphic conditions. The *Mahafale* plateau is characterised by an intermittent forest cover of dense didieracae forest and xerophilous thickets. This cover, long spared from exploitation for charcoal burning, is currently experiencing the beginnings of degradation, accelerated by the crisis, which has led to an upsurge in bloody raids to steal zebus, the mainstay of the *Mahafale* ethnic group's livelihood, coupled with several climatic accidents since 2008, resulting in a drop in agricultural production. However, far from being carried out in an anarchic fashion, a semblance of organisation in the charcoal-burning activity has been put in place, with sales outlets along the RN10, under the control of the cantonnement forestier, which states that it is preparing for the application of the new decree[116] .

In its project document, the WWF project therefore decided to set up participatory forest resource management mechanisms to better control the charcoal sector in the *Mahafale* Plateau area. These mechanisms have led to the creation of grassroots communities that will manage local natural resources rationally, with the aim of :

- protect the environment through charcoal production by disseminating improved methods and sustainable management of forest resources at village level;
- promote the local production of substitute or alternative fuels with a view to reducing wood energy consumption;
- promote new plantations to replace the consumption of natural forests, for the South-West Region, more specifically, the supply of wood fuel to the town of *Toliara*.

Five structures[117] have been set up in the *Betioky* district to manage natural resources and produce charcoal using the improved carbonisation method. These structures have been set up

---

[115] In ABETOL (Wood energy supply for the city of *Toliara*), 2008

[116] Regional Order n°022 MATD/RSO concerning the regulation of the wood energy sector in the South-West region.

[117] These structures are :

The grassroots community known as *"Mahasoa Tany"* in the *Fokontany* of *Andranotakatse* and *Tanantsoa* sud in the Commune of *Masiaboay*;

The Basic Community known as *"Ezaka"* is made up of four *Fokontany* such as *Anadabolava* and *Tsilavondrivotse* in the Commune of *Ambatry* and *Besakoa* Nord and *Besakoa* Sud in the Commune of *Masiaboay*;

The grassroots community known as *"Raza mitahy"* in the *Fokontany* of *Ereteke* in the Commune of *Beantake*;

The grassroots community known as *"Ala Mahasoa"* in the *Fokontany* of *Ampasindava* in the Commune of *Beantake*;

The grassroots community known as *"Mahavitazy"* in the Fokontany of *Anjambalo* in the Commune of *Beantake*.

with specifications and operating quotas. It has also set up a self-financing, decentralised system for monitoring charcoal production and flows. More specifically, management transfer strategies should take account of the following recommendations:

- Broaden the range of wood species that can be carbonised by downgrading certain species;
- set an annual production quota, taking into account :
    - forestry opportunities ;
    - the annual growth of the stands ;
    - the annual increase in applications ;
- not to exceed the 50% downgrading rate so as not to damage the richness of the flora and biodiversity.

In order to make the carbonisation process viable and ecologically acceptable, WWF has decided that the downgrading of species destined for carbonisation is inevitable.

6.4.2.1 CoBa and the internal control circuit :

For internal control, an organisation has been established between the producer members and the controllers or "forestry agents" per CoBa. Firstly, only members on the list of registered charcoal-makers can produce charcoal; they must have a producer's book. Before the cooked coals are bagged, the product must be checked by the inspectors. During the inspection, the forestry officers must see the kiln and the products. Once they have seen the kiln and the products, they check the name of the producer, his registration number and the number of bags against the number on the list. The forestry officer signs the production book after the check. The producer then goes before the president of the CoBa (who holds the coupons) to pay the price of the coupon, since a coupon is worth 750 *Ariary* for a 40 kg bag of charcoal. Only the president can write on the coupon, and he also signs the producer's booklet. After that, the producer presents himself and brings his products to the depot with a product booklet. The depot manager in turn checks the carnet number and waits for the depot manager to make the sale.

The consumer buys from the depot and pays the rebate to the Commune. The Commune in turn issues a receipt for 250 *Ariary* per bag with an authorisation to leave.

Before issuing a pass , the head of the cantonment compares the number of bags on the coupon with the number on the receipt.

The producer pays money to the chairman (750 Ar per coupon) and the consumer outside the Commune pays a rebate of 250Ar.

The president checks the register to see the number of the charcoal maker, the number of charcoal produced and the verification made by the forestry officer. He pays the coupon money to the treasurer. In theory, the money received is divided as follows:

- management fee: 80 Ar ;
- coupon : 40 Ar per bag ;
- the share of forestry agents or inspectors: 50 Ar per bag;

- the share of the Cantonnement forestier : 150 Ar per bag ;
- DREF forest duty: 430 Ar per bag

6.4.2.2 Training on the improved carbonisation method

Before implementing the policy to ensure the sustainability of forest resources on the *Mahafale* plateau, and to better control the traceability of charcoal, training in improved charcoal-making is essential. Charcoal burners are trained in the improved charcoal technique so that they abandon traditional charcoal burning.

After receiving training in improved charcoal burning, in order to reinforce or observe at first hand how the depots are managed, but above all so that they can take ownership of the new charcoal-burning method, an exchange visit to *Mahajanga* was organised by WWF. The charcoal burners visited GESFORCOM project sites where the improved carbonisation method had already been used. The charcoal burners were accompanied by the *Betioky* district chief and the mayors of the three communes.

On arrival in the *Mahafale* plateau, the local communities and authorities, more specifically the town councils, are ready to regulate the exploitation of this sector.

The Commune is therefore issuing a municipal decree to set up a sales depot and a control barrier, appointing a municipal agent and demarcating the coal market.

As a result, the CoBa and the Commune each have a very important responsibility in the implementation of the control, whether at the level of each CoBa or at the control gate.

However, the sales depots have changed while most of the coal merchants have not taken ownership. As a result, the new depots are not supplied. It is difficult to apply the new regulations on the operation of the coal industry because the structure, in particular the management committees of this CoBa structure, are not professional charcoal makers. They cannot ensure the internal control of: the practice of improved carbonisation, the conservation zone, the cutting of wood in usage rights, bush fires and the movement of people in the forest.

It should be pointed out that there are three stages of control, such as verification at CoBa terroir level by forestry agents; control of papers (coupon i.e. CoBa quota and pass) by the communal agent and mixed control by the gendarmerie, the Commune and the forestry cantonment at the control gate on the RN10.

Three months after the implementation of this policy, those in charge of control were unable to fulfil their responsibilities, so they stopped control and the charcoal-makers returned to their traditional charcoal-making practices.

Conclusion part 2

The second part enables us to analyse the local organisation and the synergy of the participation of all the stakeholders in the conservation of forest ecosystems, in order to detect the conditions and the degree of their participation in the management of the forest resources of the *Mahafale* plateau. It emerged from these analyses that the financial autonomy of government departments and the CoBa, the level of education of communities, land tenure problems, **the narrowness of** user rights and climatic shocks had a major influence on the motivation of stakeholders to participate in conservation. This was illustrated by the way in which government departments and local players are organised at community level, and by the dynamism of women.

In addition, the analysis of vulnerability highlights the diversity of stakeholders and their functions, which do not react in the same way to the changes introduced by the transfer of management. The analysis of interactions enables us to understand how management transfers can, over and above their objectives, induce pressure on certain functions or, on the contrary, stimulate ripple effects likely to bring about changes in the management system.

We can say that the use of the participatory approach as an intervention methodology in the protection of forest ecosystems is interesting, because it responds to the concern of donors and local authorities to involve grassroots communities closely in the development of their territory. For this reason, we felt it necessary to analyse and criticise the participation of grassroots communities in the following section.

Part Three: Analysis and criticism of grassroots community participation

To legitimise their action, conservation projects try to improve the participation rate of the target population. To do this, environmental NGOs use the MARP method (Accelerated Participatory Research Method) to initiate participation.

Moreover, international donors have asked for democracy to become the framework guaranteeing their investment, as it goes hand in hand with "participatory development".

Democracy is seen as an appropriate means of stimulating village participation in the management and control of public affairs, including the conservation and/or management of renewable natural resources. The introduction of democracy and decentralisation should facilitate the programme's action of involving the rural population through local players. However, this democracy alone cannot encourage the participation of all the inhabitants, as other elements must come into play.

But what is it that most influences villagers' motivation to get involved in conserving forest ecosystems? And what are the conditions that motivate citizens to take action for the development of their locality?

In the following chapters, an in-depth analysis will be given, followed by criticisms **of** the conditions for community participation in forest resource management and participatory development in rural areas.

Seventh Chapter:

The scope and limits of the approach itself

Why is it that poor countries only experience failure in the development policies they adopt? This is the question that has arisen.

In principle, these poor countries are characterised, on the one hand, by economic disarticulation resulting in a hypertrophy of the tertiary sector accompanied by accelerated urbanisation. On the other hand, they are characterised by insufficient production. The latter is unable to satisfy the most essential needs of the entire population. This problem leads to famine, shortages and epidemics at community level. Insufficient production results in low household income and a low development indicator. These poor countries are dependent on external aid from international bodies such as the World Bank and the IMF, through stabilisation and restructuring policies.

This external intervention was doomed to failure given the explosion of the debt crisis of the 1980s. This crisis gave rise to the concept of structural adjustment policy. All the development policies advocated by poor countries should refer to this structural adjustment policy.

Development policy is an extension of public action outside the original field of intervention. When we talk about development policy, we are referring to development institutions, which are *"a set of complex interactions between macroeconomic think tanks and decision-making bodies, bureaucracies and administrations, groups and social actors"*. This development policy emphasises state intervention to correct imperfections and promote economic growth and even development.

Poor countries such as Madagascar, faced with their poverty, try to adopt policies to resolve temporary problems known as "cyclical", or to restructure the economy according to the overall objective of these policies or "structural policy".

Malagasy experts became aware of the problem as a result of the repeated failures of strictly economic models. At the time, different models of approach had existed one after the other during the implementation of development strategies in Madagascar. First there was an integrated approach, then a concerted approach, then a participatory approach and now agar. In the end, Agar is considered to be the only approach that can be adapted to Madagascar, as it gives a voice to communities.

So, you need to know that Gelose is ultimately a new and extremely original project which consists of getting the local populations to sign a contract with the State. Signing a contract with the State means committing to a certain number of actions to protect their natural environment.

If Madagascar's biodiversity is being considered by researchers and donors, it is because since the early 1990s and the emergence of the III^{è} Republic in 1992, the country has been emerging very slowly from a long period of closure, which contributed to the deterioration of its infrastructure and the living standards of its inhabitants.

Considered to be one of the poorest countries on the planet, but also one of the richest in terms of specific biodiversity, Madagascar has for the past two decades been the focus of a series of development measures designed to meet the challenge of combating poverty while protecting biodiversity.

7.1 Scope and limits of legal instruments

The 1990s were a period of socio-political upheaval in Madagascar, culminating in the adoption of a general policy based on liberalisation of the economy, disengagement of the State from the productive sector and decentralisation. In terms of natural resource management, the following three facts have been observed:

- Due to a lack of resources of all kinds, the authorities in charge of this activity are increasingly unable to stem the tide of destruction and illegal exploitation of these national assets, because of the existence of open access situations;
- People living close to resources are both victims and, to a large extent, responsible for the spiral of degradation of Madagascar's natural environment;
- village communities apply certain forms of traditional resource management that are not valued because of the exclusive system applied by the authorities in this area.

The new agar-based approach has been in use in Madagascar since 1996. Law 96-025 governs this approach. The approach is copied from the colonial management of the Malagasy forest, whereas the local communities have long had a traditional form of management. The latter favours free access by local people to forest resources. Experts and donors are now trying to replace this traditional management with the famous Gelose or GCF decree. This leads us to identify the limits and successes of the latter, as well as the limits of management transfer.

7.1.1 The strengths and weaknesses of agar and the GCF decree

The promulgation of the Environment Charter in 1990 was the starting point for a change in the State's idea of the strategy for managing renewable natural resources, which until then had consisted of giving the administration exclusive responsibility for this management. This period in the 90s also coincided with a series of socio-political upheavals in the country, culminating in the adoption of a general policy based on the liberalisation of the economy, the disengagement of the State from the productive sector and decentralisation.

With regard to the management of natural resources, there are three main observations: the administrations in charge of this activity are increasingly powerless to stem the tide of destruction and illegal exploitation of these national assets, due to the existence of open access situations for lack of means of all kinds.

Finally, village communities in several parts of the island apply certain forms of traditional resource management that are not exploited because of the system of exclusivity applied by the authorities in this area.

In an attempt to find solutions to these problems, the Malagasy government, with the support of donors, has set up a project for the secure local management of natural resources. This is a patrimonial management method developed in 1996 to give the people of a village back control over the ecological conditions in which they live. They believe that this project restores the communities' right to have their say.

Above all, Agar is an extremely original project, one of the only ones to have attempted to put people at the centre of their own development. It is also fundamentally a non-directive approach, and some people are campaigning for a return to more dirigiste management. Agar needs to be evaluated. However, before we can evaluate Agar, we need to wait for a sociological generation, i.e. 25 years of implementation, because it has a long-term impact, otherwise we won't be able to draw up an assessment.

However, it's not even 25 years yet, and we'd like to know how we've got on with using agar. We'd like to know whether we're being treated as guinea pigs and whether we've wasted any time.

20 years have already passed since it was implemented in Madagascar. So what can we say today? What assessment can we make?

After 20 years, of course, implementation difficulties are mainly caused by interface problems. It's not the project that's causing problems, and it's not the people either. It's a problem of the implementation interface.

Opponents say that 20 years is enough to make a negative assessment and to stop wasting time trying to ask people what they want, what they dream of, what they want as a society. Stop asking these questions and stop everything because it's not getting anywhere. Taking away the voice of local residents or imposing it on them again is a major step backwards. Imposing a project on them because we want to make something for them without their consent is also a waste of time because we're in a hurry.

However, after 20 years, the strengths and weaknesses of Agar and the GCF decree have been identified.

First of all, the strengths of Gelose are as follows: it is a mother law for the transfer of management of renewable natural resources. It is cross-cutting, as it applies to all resources whose management can be transferred. It is consistent with development policies and programmes, as well as the legal framework for the transfer of natural resource management. Finally, it sets out the procedure for implementing the transfer of management: the resources concerned, the players, the process, the conclusion and termination.

The GCF decree is governed by decree 2001-122 of 14 February 2000. Compared with the Gélose decree, the GCF decree is much less complex: it does not involve either the mediator or relative land security (SFR). Only two signatures instead of three. In fact, the Mayor will not be a signatory, without this personality being sidelined as some critics claim: assessment of the capacity of the COBA (article 4), receipt of the application and its transmission after opinion, member of the local commission of enquiry (article 10), recipient of reports, P.V of offences originating from the forestry administration (article 29), communal forest law enforcement officers involved in the control of offences, empowered to seize offending products (article 30), intervention in conflict resolution (article 36 and following).

In addition, the GCF decree is more pragmatic than the Gelose decree. It is a technical management instrument, specified in terms of a management plan, the management method, whether on a self-managed or sub-contracted basis, and tendering.

However, in the 20 years since they were first used, weaknesses in these approaches (agar and GCF) have been discovered.

In the case of agar, we have noted that the application texts have not been applied to :

- the conditions for drawing up the list of resources likely to be the subject of a management transfer, by regulation (article 9 of the law); this would be an interministerial decree that has already been drawn up but not published;
- the content of the application, which must follow a standard form laid down by regulation (Article 10 of the Act);
- the content of the application must follow a standard form, to be laid down by regulation (Article 14 of the Act);
- the institution of benefits, by law, for the benefit of COBA, for the marketing and development of natural resources or derived products, (article 54 of the law);
- adjustment of these benefits to market economy conditions, by regulation, also article 54 of the law.

In addition, Gelose has experienced problems of funding in terms of approval and contracts, mediation and the mediator, and relative land security.

As far as the substantive problem with the contract and approval is concerned, the latter is a duplicate of the contract, which only makes the system more cumbersome. It is issued after the parties have committed themselves by accepting and signing the administrative act conferring management on the grassroots community, i.e. the contract.

If it were a simple agreement in the form of a visa affixed to the request from the parent municipality, or a simple letter of no objection, allowing the contract to be concluded, approval would make sense. On the other hand, everything is judged subsequently, in relation to the contract: withdrawal of approval in the event of non-performance of the obligations laid down in the contract, by the CoBa, renewal.

The identity of the competent administrative authority that issues the approval is not clear; the first tendency is to think of the District chief or his representative at the communal level; but when article 4 of the Gelose law states: *"This competence is determined by the applicable laws and regulations according to the category of membership and the nature of the resources in question"*, it is obvious that this competence is variable but it cannot be the District chief. Under these conditions, it would be the representative of the technical department responsible for the management of the resource in question; only, this representative of the technical department managing the resource is a signatory to the transfer contract; this once again confirms the uselessness of approval in its current form.

For mediation and the mediator, the transfer of natural resource management involves several entities:

- the population living alongside the natural resources, which is not already homogeneous (population from different caste hamlets) and which is going to form an association to manage a common asset for the common good;
- the Commune, which has its own vision for economic development and environmental conservation;
- the technical department responsible for managing the resource, which implements the State's policy on resource management, with an obligation to achieve results;

- technical and financial support organisations, which are essential given the country's technical and financial capacity, and which have their own imperatives and requirements.

Bringing all these people together necessarily requires mediation, which can be carried out by people from prestigious schools, institutes and universities. The problem with Agar is that it tries to create a specific profession of environmental mediators.

Finally, as regards relative land tenure security, the Gélose Act makes no mention of land tenure security in its provisions. Its introduction only adds to the complexity of the system. If there really is a need for relative land tenure security, as may be the case, it should be made optional.

As with agar, there are also implementing regulations that have not been included in the GCF decree:

- for the application order relating to the model GCF contract (article 9 of the decree), the model is annexed to the decree but no order confirms it;
- for the model specifications for FCM, the approval order provided for in article 13 of the decree has never been issued;
- the feecalculation method provided for in Article 16 of the Decree has not been implemented;
- The model subcontracting agreement, setting out the rights and obligations of CoBa and the approved forestry company (article 21), has not been published;
- Article 25 of the decree setting the rates, collection procedures and distribution of fees in the event of subcontracting has not yet been published.

Fundamental problems were also discovered: inconsistency with the Agar law and its implementing texts (GCF decree):

- the first paragraph of article 2 of the GCF decree clearly states that contractual forest management falls within the framework of the objectives and prescriptions of the Gelose law; expressed in this way, the GCF decree is truly inconsistent with the Gelose law; the term "prescription" is too much, as it implies the application of all the provisions of the law, in particular the intervention of the mediator and the signature of the mayor on the contract;
- The question that arises is whether this legal inconsistency has had any impact on the effectiveness of the GCF contracts. It seems not, because in fact, even though the mayor is not a signatory to the contract, he is fully involved in the process of setting up and implementing the management transfer. As far as mediation without a mediator is concerned, every precaution seems to be taken to resolve the emergence of conflicts, which cannot, moreover, be 100% avoided.

The tools prescribed for the GCF decree have not been designed. For example, the system for monitoring and evaluating the contract after three years of implementation requires a procedure to be put in place. This is provided for in the text, but the administration responsible has not considered it essential to draw it up. How will the people responsible for evaluation go about it? If the central administration had designed the procedure and distributed it to the decentralised level, the latter would have applied it and there would have been harmonisation throughout the country. Fortunately, a guide to the monitoring and evaluation of natural resource management transfer contracts has now been designed and circulated. The first thing that needs to be done is for the forestry administration, in collaboration with its management transfer partners, to field-test the guide on GCF contracts and draw up a monitoring and evaluation procedure tailored to forest resources.

7.1.2 Limits of the Management Transfer

Forests are at the heart of natural resource management policies, as they are the source of many products needed by the Malagasy people, such as building materials, domestic needs and medicinal plants. It also ensures that production factors are in good condition, protects the soil against erosion and regulates the water regime.

We can therefore reiterate that the traditional way of managing Madagascar's forests defends free access to natural resources, and this measure has repercussions on the sustainability of the use of these resources. Forest resources have become a resource for a population faced with energy, food and construction problems on a daily basis.

The growth in the economic situation promised by the current President of the Republic has yet to be felt by the great mass of the population, particularly the peasant farmers. Indeed, it is against a backdrop of impoverishment that the trade in forest products, which is firmly rooted in the informal sector, helps to generate the cash resources that are vital to the survival of a large part of the population.

The socio-economic and political environment of the forestry sector is, however, quite fragile insofar as the forest is an ideal arena for the deregulation of political, social and economic systems. When one of these systems malfunctions, it is the forest resource that suffers the consequences, in the form of an increase or decrease in the number and intensity of forestry practices.

Moreover, for several years now, there has been a serious deterioration in forest resources. This is due to over-exploitation for timber and energy production, the maintenance of extensive agricultural practices (clearing and bush fires) and demographic pressure, which indirectly accelerates the process.

Degradation of forest resources leads to degradation of the agrarian system by affecting irrigation systems and reducing agricultural production (through erosion). This degradation leads to a number of economic and social problems (lower agricultural yields, rural exodus, increased crop theft, etc.), which affect the development of rural areas.

The *Mahafale* region is increasingly affected by the phenomenon of desertification linked to climate change due to the irregular distribution of rainfall. For more than a decade, this problem has resulted in both vegetation degradation and soil degradation in the cultivated areas of the plateau, the most significant of which are found in the coastal areas (Communes of *Beheloke*, *Itampolo* and *Soalara* Sud)[118] . 55% of the population can no longer meet their minimum energy needs[119] , and most of the population has become poorer in recent years [Barraud, 2006].

To solve these problems, the authorities are implementing a new natural resource management policy that places greater emphasis on decentralised territorial collectivities, more specifically

[118] FOFIFA / INSTAT / Cornell - 2001

[119] In Projet COGESFOR du plateau Mahafale : Rapport d'activités annuel (Mars-Décembre 2010)

the communes. This new management method is part of the Gelose approach, in which the management of local resources is transferred to local communities or *"Vondron'Olona Ifotony"*. However, this sustainable management of forests by local grassroots communities has long-term impacts. The transfer of management to these communities has its limits.

7.1.2.1 The level of education of the community and the management committee

Even if agar and the GCF decree are the management methods that give most responsibility to the communities, they are a shared management between the communities and the State, because renewable natural resources belong to the State. The main player in the management of natural resources is therefore mankind, and more specifically the grassroots community concerned. Through this management method, local populations begin to exist in a concrete way; their heterogeneity and complexity are revealed, and they are imaginative and concerned about their own future.

Managing forest resources under the Agar or GCF contract requires a minimum of skill and knowledge of the management techniques that correspond to this adopted model.

Historically, there has been no proven sustainable management system, and above all, there is no precise definition of sustainable management of natural forests. In other words, there is no perfect, immutable model for managing forest resources that communities can emulate, but the management of forest resources depends on the management plan and specifications required for a transfer.

The forestry administration most often requires a detailed management plan before granting a management transfer. Drawing up a detailed management plan is a highly technical task, given that the communities have a very low level of education and are even illiterate. As a result, they are unable to draw up these plans on their own because of their incompetence. So the technicians from the environmental operators are going to draw up the plan instead of the grassroots communities. The latter have not mastered the theories set out in the document. The technicians have not written down the management policies desired by the communities, but they have drawn up the content of the document in a more technical manner. The communities are illiterate and find themselves at an impasse when it comes to implementing the document. They cannot undertake the activities set out in their management tools in the desired direction.

In short, the illiteracy of the local communities is holding back the implementation of the management plans and compliance with the commitments they signed up to in their contracts. This inability on the part of the grassroots communities is preventing them from properly mastering the new management, even though the success of this management and the promotion of a well-protected environment free from all forms of degradation is a necessary condition for the sustainable development of the region in Madagascar. The illiteracy of natural resource managers creates a gap in the forest resource management policy, which leads to its implementation in a direction that the forestry administration does not want. They can also cause problems for the technicians who provide the training needed to implement the transfer, as they are not familiar with the current situation. Hence the unsuitability of the techniques used.

7.1.2.2 Tension between indigenous groups and migrants

Setting up a structure to manage forestry resources will ensure that the policy of sustainable management of renewable natural resources is implemented.

In the case of management transfer, the structure known as the grassroots community is preferred for implementing this sustainable management policy. The principle of this grassroots community managing renewable natural resources is different from that of the *fokonolona*, which is the social structure based on Malagasy values. The *fokonolona* is the most stable, solid and durable structure, accepted by all, especially when the country is going through a crisis. As for the grassroots community that manages the forest resources, it is heterogeneous.

> *"In their documents and in the way they negotiate with local authorities, public authorities (including NGOs) favour the notion of community ("basic community" ...), even though this is far from being a reality. A concrete basic social unit (a group of lineages, allied or not, living in a restricted area, united by a dense network of daily relationships fostered by proximity) in no way gives the homogenous image evoked by the notion of community. The idea of a community grouping together various social elements under a single pyramidal authority ready to defend the uniform concerns of its members is particularly inaccurate. In these pseudo-communities, we find several authorities that are poorly articulated with each other (the chiefs of the main lineages, a few rich herd owners, etc.) who are far from defending the same objective because of rivalries between lineages engaged in contradictory strategies to win the race for ostentation, rivalries between mpanarivo engaged in struggles, often merciless, for local hegemony, conflicts of power, particularly over land, between natives (tompontany) and older migrants who have sometimes been settled for several generations...".* [Chantal Blanc-PAMARD & Emmanuel FAUROUX, 2004].

Because of the intense impoverishment of rural Madagascar, members of the community aspire to something more beneficial in the short term, which has a direct impact on their survival needs. Farmers see only one priority: the immediate resolution of their shortages. However, all the important natural resources, mainly forests or lakes, that can be the subject of an agar or GCF contract have long been used by several social groups or several communities.

Natives, for example, who are familiar with the local spirits and are able to derive various benefits from their role as intercessors with these spirits, tend to be much more protective of the environment than migrants. Some villagers, especially those involved in transhumance, wish to conserve the forest that provides refuge for their zebus in certain circumstances. They want to keep out other immigrant groups, especially those with long experience of managing forest resources. They want to control these resources on their own because they believe that they belong to their ancestors.

On the other hand, the members of the community who are most vulnerable because of food security continue to clear the land. On the other hand, migrants, both old and new, try to invade territories, and this offensive will lead to communities losing control of their resources. This rivalry between migrants and natives therefore leads to latent but sometimes overt conflict.

By way of illustration, various hamlets have been discovered within the CoBa de *Vorojà* territories. These hamlets are being created little by little over time, so it is difficult to say exactly how many there are at the moment. The inhabitants of these hamlets are all sedentary immigrants. They try to exploit the forest resources.

There is also a conflict of interests between the members of the community: for some, it is in their interest to protect the resources, because they are the source of aid and funding, and for others, especially the migrants, it is in their interest to try to erase the image of the farmers as locked into the immutable, unfit for progress, by clearing land to maximise their income at a time of high maize prices, or by marketing the most sought-after forest resources.

In addition, the beginning of the loss of respect for traditions and elders shown by young people influenced by the city is generating social conflict and governance problems within the CoBa. This conflict weakens the sustainable management of renewable natural resources. In addition, the insecurity prevailing in the region has influenced the communities not to fully exercise their responsibility for natural resource management, as those who have lost their interests have taken advantage of this situation and consequently manoeuvred to prevent the proper application of the development plans and commitments enshrined in the CoBa management tool.

Finally, the balance of power between natives and foreigners, whether overt or latent, hinders the implementation of the management transfer, especially when the foreigners *(mpiavy)* try to control the territories and forest resources. With the CoBa unable to patrol forest resources, land clearers and timber traders try to take advantage of the situation to do business.

7.1.2.3 Non-ownership by community members

Most of the CoBa set up in Madagascar received awareness training from environmental operators before committing themselves to managing renewable natural resources. They were not familiar with this shared management method, which is the Gelose or GCF decree. No member of the community opposed the implementation of the management transfer; no one refused to comply with the provisions of the *dina* and no villagers questioned CoBa as the manager of these forest resources. However, the majority of villagers are not members of this CoBa. As a result, the members of the CoBa are far from representative. It seems that the application of the tool through the formal establishment of community associations called "basic communities", but which exclude certain members of the community, sometimes leads to abuses. This exclusion has led to a *lack of responsibility on* the part of certain community members in the management of resources.

In order to exclude some villagers, board members are sometimes unwilling to talk clearly about the conditions of membership. What's more, the CoBa management committees never have an opportunity to present the existence of this structure, which manages local forest resources, to the entire community, unless they organise a rite of thanks to the spirits who own the forest.

As a result, the communities know very little about the management of these natural resources and leave the CoBa alone to protect the forest, believing that the forest belongs to the environmental operators and that the CoBa are merely "guardians". Taking advantage of this situation, rural stakeholders are not involved in the CoBa and derive little income from the exploitation of forest resources. As a result, even if community members find infractions in the forests, they will not report them to the management committee because it is none of their business.

7.1.2.4 The non-application of the *dina*

Traditional Malagasy society is a lineage-based society. It does not spontaneously function in a democratic manner and often has difficulty managing problems that go beyond this lineage framework. However, Malagasy technicians, through the State and donor agencies, have worked with this traditional structure to put in place the Gélose law and the GCF decree to ensure the shared management of forest resources.

As mentioned in the previous paragraph, the structure put in place to manage natural resources seems strange, as it in no way conveys the homogeneous image evoked by the notion of a traditional Malagasy community. Many villagers (whether members of the CoBa or not) were not involved in drawing up the *dina* or the management plan. They do not know much about the content of the *dina*, even if they were not opposed to the CoBa organisations.

Moreover, the insurmountable obstacle in the application of this *dina* is *fihavanana.*[120] With other neighbouring communities, if lake habitats and grazing lands are transferred, there is a risk that they will be destroyed. For several generations, other communities and especially the *mpanarivo* of the village have been in the habit of taking their herds of zebus to the pastures of *Andremba* (the arrival zone for transhumants from the northern coast) during the transhumance period. In the lakes, neighbouring communities have free access, as water is the main problem in the region.

The sense of *fihavanana* is another reality to be managed within the communities. Some CoBa continue to issue logging permits to other neighbouring communities, despite the provisions of their management instruments which, incidentally, we doubt the CoBa read. Whenever we alluded to these cutting permits being issued to other communities, our interlocutors always replied that, in the name of *fihavanana*, it is difficult, even inconceivable, for a CoBa to deny neighbouring communities access to certain species that they do not have. This is how, for example, the CoBa of management transfers such as *Reniala* d'*Ampotake*, *Mitsinjo taranake* d'*Itomboina* and *Mahasoa* d'*Ampitanake* have agreed to allow neighbouring communities access to their resources.

To resolve a social problem, debates lead to collective agreements, *titike*[121] or *dina*, which are binding on the co-contracting parties, as the agreement is celebrated by a sacrifice of oxen. The ancestors thus become demanding witnesses. Disobeying the agreement would therefore be as serious a fault as disobeying the ancestors. On the other hand, in the case of the regulation of offences in the forest, the *vonodina* is generally made of money and the witnesses are the administrative authorities but not the ancestors.

In addition, the proliferation of insecurity in their society means that communities are afraid to exercise their responsibility for managing natural resources. Indeed, if communities are unable to settle cases of *dina* offences between themselves, they turn to the legal authorities (territorial authorities, law enforcement agencies, courts), which may leave offenders free and unpunished. Those who fail to comply tend to re-offend and even take revenge, while making a mockery of social contracts, which have little value before the law.

---

[120] *Fihavanana* is the foundation of Malagasy culture. *Fihavanana* translates into mechanisms of solidarity, support and unity at both family and community level. The spirit plays a key role in *fihavanana*, as the Malagasy believe that it is the *fanahy* (spirit) that makes a man.

In passing, we should also point out the difference between damaged *fihavanana* and destroyed *fihavanana*. The difference is that damaged *fihavanana* can be repaired through repentance or a fine, albeit with the observance of a rite. In the second case, the relationship is totally destroyed, the *fihavanana* is dead. In a way, this is the divide between a fault that can be repaired and a fault that cannot be repaired. It is this fear of destroying the *fihavanana* that sometimes leads a community, in spite of itself, to make a concession to its neighbours.

[121] *Titike* = ceremony designed to make a collective agreement or *dina* more solemn, but concerning a geographical area that is generally more restricted than the *dina*.

7.1.2.5 The degree of involvement of the forestry administration

CoBa's essential mission is to ensure the sustainable management of natural resources, in accordance with the management transfer contract, by regulating their access and use. Rules are only effective if they are accepted by users. User acceptance depends on their level of knowledge and understanding of the rules. It is up to the forestry administration to support the communities that manage natural resources through capacity building.

Our observations, backed up by the qualitative analysis of the focus groups, revealed that the technical forestry services are still too often not involved from the outset of the transfer (in some cases, they are only informed by the support organisations very shortly before the contract is signed). However, it should be pointed out that, as the State's representative at local level, the cantonnement forestier has full authority over the forests in its area before the management transfer contract is signed. So it is part of the power of the Cantonment forestier that is being transferred to the CoBa. Before the management transfer system was put in place, towns were mainly supplied with wood from forests managed by the decentralised forestry services.

During the implementation phase of the management transfer, it is up to the forest administration or a forest district as signatory of the transfer to carry out the controls as well as the periodic monitoring of the CoBa with the resources they manage. Moreover, the role of the forestry administration in the implementation of the transfer of forest resource management is to carry out monitoring and control. To do this, it can inform the CoBa of new rights concerning the management of natural resources, train members and provide technical support for the association; take decisions concerning the misconduct of CoBa members, and apply the corresponding penalties, or even dissolve the contract in the event of an infraction.

However, due to very limited financial and human resources, the forestry administration is not in a position to fulfil its roles and ensure that the conditions for exploitation comply with basic management rules as set out in forestry legislation.

In addition, the mining policy adopted by the current central government also threatens the conservation and management of forest resources by communities. The state authorises mining or industrial or tourist investments without consulting or obtaining the consent of the local communities. This is all the more true given that the government wanted to activate the demarcation of protected areas following the presidential decision in Australia in 2014, in order to be able to process mining permits. However, the communities do not yet have all the *"weapons"* (technical, legal, etc.) to defend their interests, even in the context of a public consultation on an environmental impact assessment, in the event of a threat to the destruction of their resources. The other forest districts are taking advantage of this situation by having a bad time.

Under these conditions, the *Tsimanampesotse* National Park, which borders the *Mahafale* plateau to the north and forms the hard core of the landscape with its 207,000 ha surface area, equivalent to 12% of the total surface area, as well as the management transfers, are exposed to various dangers, particularly those of anthropogenic origin. The most characteristic of these are the "*teteke / hatsake*" or slash-and-burn cultivation practices and transhumance practices, which tend to be long both in duration and distance. These practices lead to overgrazing, degradation of the plant cover and soil, and social conflicts between farmers and herders. This situation is exacerbated by the continuing increase in population numbers.

In view of the insufficient effectiveness of the regulatory and legislative measures banning land clearing and deforestation enacted by the forestry administration, or the absence of appropriate rules for the conduct of transhumance, combined with the precariousness of living conditions in this region, local people cannot help but resort to practices that do not respect the environment in order to survive. As a result, the integrity of the *Tsimanampesotse* Park and, in particular, its outlying areas, has been compromised and the local people continue to live in persistent poverty.

As a result, the ecological crisis in the areas of arrival and the saturation of land in the areas where transhumant herders are received have led to a considerable increase in the phenomenon of intrusion into natural resource management transfers and into the new areas of the park. Livestock breeders have taken advantage of the *Malaso* phenomenon and the[122] agreement between the communities and the park authorities on the use of the park area, which was previously considered unhealthy. In addition, as the resources made available (tracks, carts, etc.) for the surveillance of the forests transferred to the communities have so far been clearly inadequate (especially in the pastoral reception zones for those that have them), there were few obstacles to the herdsman's desire to carry out his transhumance within the *Tsimanampesotse* park.

Clearing land or *teteke*[123] is a traditional land development practice already used by the first occupants of the *Mahafale* plateau. At the time, the low population levels made it possible to combine the practice of clearing land with the use of the forest. At the time, the forests had a sacred character (tombs were generally placed there) and constituted a reservoir of food and medicinal plants for people and livestock. People used to take their food from the forest, but when the forest was destroyed, they were forced to start clearing land to support themselves. However, the original thick xerophilous thicket has not disappeared everywhere, although it has become impoverished and transformed into *monto* (open secondary thicket) and *sarike* (closed secondary thicket) in many areas.

Thus, the new migration dynamics in the *Mahafale* plateau have emerged as a determining factor in the forest degradation currently observed in the region. While immigration is not a recent phenomenon in the south-west region, it has changed radically over the last 15 years or so. In the past, migration was temporary and individual; today it has become massive and permanent.

---

[122] The local population and the MNP refer to it as the "Itampolo Ten Commandments". The second article of this agreement stipulates that ANGAP will demarcate a grazing area for transhumants. As a result, the pastoralists of Itampolo can transhumance there.

[123] The local word *teteke or hatsake* refers to the action of cutting down and burning the trees directly, in order to sow without any further preparation of the plot. This system is practised on virtually any terrain, including the stony, shallow soils of the limestone plateau, as it mobilises the nutrients accumulated in the litter and forest biomass. Cleared plots are cultivated, generally for two years, then abandoned because of a drop in fertility and/or an invasion by weeds that would make it necessary to invest a large amount of labour in weeding. New land is then cleared for a new crop cycle. After this abandonment, a secondary forest (*moka*) forms again after about 5 years. The field can be recultivated after 10 years, but with much lower productivity.
This choice of clearing is explained by the ease with which the forest can be cleared and then cultivated, compared with the difficulty of preparing the soil and weeding older fields. Similarly, the soil in woodland areas is more loose and fertile, making it more suitable for cultivation, compared with the hard soil in savannah areas.
Land clearing is therefore an extremely productive technique, both in terms of the area cultivated (3-5 hectares of land clearing can be cultivated per family per year with family labour alone) and in terms of the labour invested.

Gradually, the migrants became richer, and their social cohesion was consolidated, whereas the natives did not benefit from political opportunities enabling them to acquire land titles. As a result, they embarked on a "conquest" of the forest. On the one hand, in economic terms, they exploit the environment (charcoal production, tree felling, dry slash-and-burn farming). Secondly, in symbolic terms, they have "repopulated" certain areas of the forest with spirits and deities of the *"kokolampy"* type from their territory of origin, in order to spiritually appropriate the forest area and thus control it completely.

Following Madagascar's independence, many families from the coast began to settle inland on the plateau. The dynamism of slash-and-burn farming was intensified at the same time as the *Vatoeka*[124] (state agents) began to buy local produce from farmers. The rise in the price of maize attracted some farmers to cultivate the land and exploit the forest. What's more, in the 2nd Republic, obtaininga fire permit from the Water and Forestry warden was interpreted as authorisation to clear the forest. As a result,from the 1970s to the early 2000s, tens of thousands of hectares of forest were destroyed to clear land. As in the case of transhumance, kinship (matrimonial relationship, blood brotherhood) plays and has played an important role in access to forest land and land clearing.

Very quickly, it became difficult for the *Mahafale* to resist the economic dynamism of the migrants, their advance into the forest and their exploitation of the forest. This loss of control over access to land and means of production is currently leading the *Mahafale* populations to react with different strategies, which are sometimes implemented in contradictory ways in the forest area: they can occupy the forest with their herds, practice deforestation themselves (slash-and-burn agriculture on the edges of management transfers or inside the *Tsimanampesotse* park), and, paradoxically, support forest protection programmes, or reinvest ancient ceremonial practices linked to the forest environment.

In the south-west of Madagascar, particularly in the *Mahafale* landscape, the forest, once a sacred and marginal place, has become a key space where new socio-economic issues are expressed. Rivalry over control and exploitation of the forest environment is the materialization of tensions and conflicts between indigenous and migrant populations.

The main reasons given for the increase in land clearance are as follows:

- the arrival of new occupants in the *Mahafale* plateau area who have appropriated new land;
- the return of pastoralists to agriculture due to the *malaso* phenomenon;
- the introduction of parks and TGRN, which has pushed coastal dwellers into undeveloped and unprotected areas (Kaufmann JC, Tsirahamba S., 2006);
- the rise in the price of maize and the ever-increasing number of collectors;
- the major droughts that have hit the area (the clearing front, for example, advanced considerably on the cliff after the drought of 1990-1991) [Lebigre, J.M. and F. Bellera, 1997].

[124] Vaomieran'ny Toe-karena or Economic Committee

7.1.2.6 Summary of available data :

The figures available on land clearing are very disparate. According to a document produced by the WWF on the basis of a retrospective analysis of SPOT satellite images[125] , trends in land clearing in the park and the TGRN are as follows:

Table 11: Clearing trends within the TGRN and *Tsimanampesotse* park

Sites	Clearance rate 2003-2008 (%) annual	Clearance rate 2008-2010 (%) annual	Trend
Former boundary of *Tsimanampesotse* Park	0.46	0.32	↓
New extension to *Tsimanampesotse* Park	0.65	1.25	↑
Average of 11 management transfers	1.16	0.80	↓
Peripheral zone of the park (without TGRN)	1.48	2.75	↑

Source: WWF Toliara, 2013

This table shows the trend in forest clearance within the park boundary and in management transfers ofnaresources.ithin the old park boundary and the oldmanagementtransfers, the clearing front is declining. On the other hand, in the new extension of the park, which is none other than the former core zone of the TGRN, and in the peripheral zone ofthe park, there has been an increase in clearing. This explains why land clearing has always taken place on the *Mahafale* plateauand why it seems to be increasing. Most of the land clearers are members of theCoBa who are contesting the extension of the park because they believe they were cheated when their core area was taken over for the new extension.

For the average of the 11 management transfers, this decrease in clearing can be seen, but if we analyse the transfers individually, a focus of clearing is sporadically scatteredhere and there. After 2011, the area cleared in the management transfers increased both in the coastline and in*Ankara* to the east of the park. Aerial surveys carried out by WWF in 2010 and 2011 report several clearings within *Tsimanampesotse* Park (particularly in the western zone, i.e. in the new extensions).

oogle Earth images also showed the extent of land clearance on the *Mahafale* plateau, although it was not possible to date the cleared areasprecisely. Surveys and documentary analysis have shown that :

- Cleared areas are mainly located to the east of *Tsimanampesotse* Park *(Ankara)* and management transfers are not safe;

[125]WWF. 2011. *Study of deforestation in the Mahafale Plateau landscape*. Draft.

- Major clearings are also taking place within the *Tsimanampesotse* Park, particularly in the *Itampolo* sector. We have already illustrated this case using the clearing carried out by the *Ambolisogno* community in the forest managed by the *Mizakamasy* CoBa;
- some municipalities are more affected than others by land clearance, particularly *Itampolo*, *Ejeda* and *Beahitse* ;
- some transhumant herders take advantage of their presence in host areas[126] to clear land, without regard for the regulations in place;
- some of the food produced in the host areas by clearing and bulldozing is destined for the coastal zone (reverse "transhumance" of agricultural production);
- some wealthy people organised large-scale clearings by paying for labour;
- Finally, some areas used as pasture in the past are now cleared for cultivation.

Despite the positive points noted by CoBa members (regeneration of trees, increase in the number of wild animals, reduction in illegal felling, improvement in rainfall, increase in the quantity of renewable natural resources and refuge for pastoralists' livestock in protected areas and management transfers), agropastoralists are forced to clear protected areas. Indeed, they find that the fertile land on the plateau has become off-limits. In addition, communities have learned that reprehensible activities are not followed up by coercive action, despite controls on access to forests by park and CoBa agents.

The development of the country through the exploitation of its mineral resources is in vogue. In the *Betioky* district, oil exploration by a Chinese company in partnership with the Malagasy oil company PETROMAD is threatening management transfers, in particular those for wood energy management transfers (CoBa charcoal producers), as the Chinese company has destroyed the forests above the areas where it has been prospecting, regardless of the status of the area (grave, private or managed by the communities, as in the case of renewable natural resource management transfers). Similarly, the exploitation of precious stones (garnets, rose quartz, amethyst, pink beryl, granite etc.) in the *Ampanihy* district is wreaking havoc in a large part of the Etrobeke management transfer and has put a lot of pressure on the other management transfers in the District.

For the *Sakoa* coal mining project, also undertaken by a Chinese company, several management transfers are affected by the construction of an evacuation road for products as far as *Salary*, and a loading dock. It should be noted that the environmental impact study for this project was carried out by agents from the Ministry of the Environment, Water and Forestry in *Antananarivo*. In their study, they concluded that there would be no environmental impact on the transport of raw products from *Soamanonga* to *Salary* via the CoBa management transfer hubs.

The CoBa board members are protesting against the construction of the road through their core area, but the forestry administration in the South-West region has not reacted. During our visit

[126] The transhumant reception areas are located in the Ankara to the east of the park. The following CoBa are home to transhumants: *Mahasoa* d'*Ampitanake*, Tsivery *anjara Mahasoa* de *Vorojà*, *Fiarovantsoa d'Ankitekiteke, Magnasoa tane* de *Behombe*, *Magnasoa Tane mitsinjo Taranake* de *Marofototse,* Mitsinjo *Taranaka* de *Tomboina* and *Soa ho anay* d'*Andremba.*

to the site, we heard that the mining company will soon receive its mining permit and construction will begin. Is this *Sakoa* charcoal mining project in the best interest of the Malagasy nation, and can it not bypass the hard core of management transfers?

Thus, the following questions deserve to be shared so that we can reflect on how the State deals with renewable natural resources in Madagascar:

- does the government encourage grassroots communities to conserve biodiversity for foreigners?
- Is the State's encouragement to preserve biodiversity tantamount to disturbing communities so that they are fed up with occupying their territory or so that they become destabilised in their socio-economic activities and it is easy to drive them out of their homes at a time when foreign operators are interested in their ancestral land?

Whatever the answers, our interpretation of this situation is simple: the success or failure of the conservation of Madagascar's forest resources depends on the political and economic will of its leaders.

To conclude this sub-chapter, the ban on planting on the plateau in the conservation zones weakens the members of the community. The extension of the park into the CoBa conservation zones means that there is not enough arable land to the west and east of the park, as the soils there are considered to be more fertile. This extension also means that the boundary of the CoBa land is being pushed back towards the interior of their village. This retreat also means that the management committee and park officials are unlikely to take legal action, as the official cultivation areas are located to the east of the park or on the coast in the new boundary. As a result, the lack of legal action and, above all, the fragility of the decision taken by those in charge will have repercussions on the success of the transfer of management of renewable natural resources.

We can therefore emphasise that programmes to combat deforestation cannot be designed without taking into account these new issues, which are investing in Madagascar's forests and contributing to the loss of forest massifs, otherwise they will be doomed to failure. Similarly, they can only be implemented with the support and participation of local populations.

7.1.3 The effects of the management transfer

Transferring forest management to a grassroots community has consequences. These can be positive or negative. They vary according to the trend and the angle from which we have conducted the analysis. Thus, we are interested in the effects at the level of conservation and at the level of the population.

In terms of conservation, transferring forest management to the grassroots community has positive effects, especially in terms of preserving these resources, through the existence of local initiatives to combat all forms of degradation, such as bush fires, and reducing the amount of wood cut. By genuinely taking charge of the lowest levels of the community, it is possible to achieve optimum results and ensure the sustainability of ongoing local development activities.

The sustainability of development activities is based above all on the regeneration of the ecosystem, the most important (physical) capital identified as a source of growth, according to neoclassical theories. In this sense, regeneration means not only increasing production, but also increasing productivity in order to achieve economies of scale.

However, natural regeneration is not yet effective. The lack of mastery of reforestation techniques, combined with difficult climatic and soil conditions, persists and is hampering some attempts at repopulation by local communities. Madagascar has already carried out a number of transfers of natural resource management in various regions.

As far as the local population is concerned, the dynamics within the local grassroots communities can have positive effects: transferring forest management to a community effectively contributes to the revaluation of these resources. The professionalisation of this sector, with the effective marketing of products, provides jobs and regular income for the local population. Technical farmers have emerged and have perfected modern weaving techniques. Economic operators have turned to these organisations to collaborate on marketing.

In social terms, the emergence of farmer leaders from local communities has facilitated their integration into decision-making and consultation bodies. Improved relations between the local grassroots communities and the communes have provided a favourable framework for various contacts and exchanges with other technical and financial partners.

For this transfer process to work, certain conditions and principles must be respected. The question that arises is: what are the success factors for management transfers?

7.1.4 The various factors that can make or break a management transfer

7.1.4.1 Success factors

Management transfer is implemented with a view to conserving natural resources in the long term and contributing to the development of the population. Under Law 96-025, the principle of grassroots community participation in natural resource management is based on voluntary participation. But conservation and sustainable management come at a cost. The big question, still unanswered, is twofold: on the one hand, how to finance the cost of conservation and sustainable management, and on the other, what the local people, the grassroots communities (VOI) - the main players in conservation and sustainable management on the ground - stand to gain.

According to the members of the grassroots communities we spoke to, the transfer of management has had a positive impact on changing people's attitudes towards the sustainable management of resources, and on solidarity between the members of the VOI with regard to the activities to be undertaken and the decisions to be taken. This improvement is now beginning to have an impact on the forest.

By way of illustration, to demonstrate their commitment to conserving biodiversity, the representatives of 487 natural resource management communities from 17 regions of Madagascar formed the *TAFO MIHAAVO* network, which issued a declaration of commitment and proposals: the "*Anjà* Declaration" in May 2012. Since then, the network has been working to develop a proposed strategy and community action plan for the sustainable management of natural resources, highlighting the place of the VOI in relation to the *fokonolona*.

In ecological terms, the VOIs stated that since the adoption of forest resource management, a change in the ecosystem has been observed: the regeneration of forests, which has led to the return of fauna that villagers rarely found in their territory, such as the abundance of hedgehogs. Community members have also noticed a reduction or spread of human pressures. During the rainy season, they noticed an increase in green spaces. Despite this improvement, they noted the need to monitor the status of some conservation target species (the raison d'être of biodiversity conservation in the *Mahafale* plateau). These include *Cedrelopsys greveii (katrafay), Albizia greveana (lovainafy), Dyospiros perrieri (mendoravy), Chloroxylon falcatum (voaovy)* and *Neobeguia mahafaliensis (handy).* The same applies - as far as fauna is concerned - to the *Geochelone radiata* and *pixis* turtles, species *that* are subject to periodic illegal trafficking.

Figure 15: Impact of management transfer on the *Mahafale* plateau

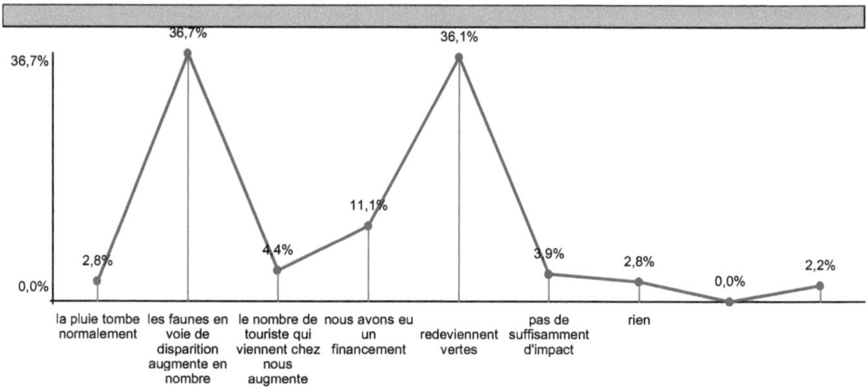

Source: Our own survey, 2014

This graph shows the results of the activities undertaken by the grassroots communities that manage forest resources. Our survey revealed that 2.8% said that, thanks to the transfer of management, the rain fell regularly; 4.4% said that there had been an increase in the number of tourists visiting their areas. 11.1% said that they had benefited from funding from the SGP/ *Tany Meva*. With regard to the ecological aspect, 36.7% of them noted an increase in the number of varieties of fauna that are in danger of extinction in their areas, and 36.1% also found that the forests in their areas were greener than before. On the other hand, 3.9% did not find sufficient impact from the transfer of management in the areas managed by the VOIs.

7.1.4.2 Blocking factors

Since its introduction in Madagascar, management transfer has been hampered by a number of factors. These can be found at the demographic, economic and institutional levels.

Firstly, in the demographic field, the degradation of the forest ecosystem in non-developed countries has accelerated in recent years due to the demographic explosion. This is also a problem for Madagascar, even though statistical data on this situation are not available.

Indeed, population growth in poor countries is significantly higher than in developed countries. On average, compared with the previous year, annual population growth in undeveloped countries is 1.83%, while in developed countries it is 0.55%[127] .

This accelerated population growth has led to a significant increase in the need to exploit natural resources to the full. The result is the destruction of these resources. This over-exploitation consists above all in seeking to satisfy consumption, energy, space and other needs.

The forest resources managed by the local communities encounter problems of intrusion by other members, whether or not they are members, in order to meet their needs, even though this transfer of management is governed by a *dina*. The application of the latter is limited by a number of needs, namely satisfying consumption, energy and space requirements.

Satisfying the consumption needs of communities around these forest resources is linked to the dependence of human activity on its natural environment. The human population is highly dependent on natural resources, especially in undeveloped countries such as Madagascar.

For example, around 53% of the population in these countries work in the agricultural sector, and the use of technology is low [Bertrand Y. A., 2007] (in the case of Madagascar, this rate is much higher). There is a positive relationship here between production and cultivated area: the more cultivated area there is, the greater the production will be, hence the need to further expand cultivated areas, as production is no longer sufficient to satisfy the food needs expressed by the population. However, on the *Mahafale* plateau, it is impossible to expand cultivated areas because of the lack of arable land[128] .

In terms of meeting energy needs, energy demand has risen sharply since 1980. This increase is due to a sharp rise in the population, which is becoming an ever greater consumer. This high demand for energy has led to the over-exploitation of resources that can supply it, especially forests, since wood remains the main source of energy for more than half the world's population. This dependence stems from the fact that the population is unable to access new techniques or

---

[127] World Bank, 1992, World Development Report, p28

[128] In MARP Report, SuLaMa 2011, research for sustainable development: "The morphology of the region is characterised by three main zones distributed from east to west and extending from north to south. To the west, there is a continuous coastal plain varying in width from 1.5 to 15 km. This area is covered by quaternary sands interspersed with alluvial deposits and temporary pools. It includes the 15-kilometre-long *Tsimanampesotse* lake, classified as a Ramsar site on 19 February 1998 and part of the national park of the same name. The lake is bordered to the east by the "*Mahafale* plateau", a nummulitic limestone plateau. This plateau, which covers an area of 10,000 square kilometres, is generally between 100 and 200 metres above sea level. It also forms a large part of the *Tsimanampesotse* National Park. It is uninhabited and displays a variety of karstic features (Raunet 1996):

- Around a hundred pot-holes 40 to 100 metres deep in the north-western part,
- An area of clay-bottomed sinkholes in the central and eastern parts,
- Depressions with decalcification clays, as well as dry corridors and valleys, evidence of a fossil hydrographic network, in the eastern part (Raunet 1996).

The decalcification clays give rise to fersiallitic soils ("red earth on a limestone shell"), which are well-structured and have a neutral or basic pH.

The third zone, located to the east of the limestone plateau, is made up of a mosaic of limestone outcrops and depressions covered by thick Pliocene sandy deposits, often referred to as "red sands". The red sands give rise to tropical ferruginous soils with a rather acidic pH ("siliceous red earth"), sometimes mixed with decalcification fersiallitic soils; or, less frequently, to more fertile hydromorphic soils in low-lying areas".

substitutes: firstly, the costs are very high, but most of the population also lacks the technical capacity to use the new technologies.

For some time now, the *Mahafale* landscape has been the potential producer of charcoal to supply the city of *Toliara*. In the report on the analysis of the wood-energy sector involved in supplying the town of *Toliara* on the *Mahafale* plateau by the NGO PARTAGE in July 2011, forests are the main, if not the only, sources of energy used by local people in the intervention area. The wood is felled to be used directly as firewood or transformed into charcoal.

Charcoal production in the area is spread out over the year, with two very distinct seasons: the dry period, from May to October, which corresponds to the off-season for agricultural work, when charcoal production is high, and the cultivation period, from November to April, when production is almost halved. Surveys in the four communes showed that 39% of charcoal burners work continuously throughout the year; 23% produce charcoal during the off-season and 38% do so occasionally. The total quantity of charcoal delivered to the town of *Toliara* is estimated at 397 tonnes per year. WWF and the forestry administration have tried to organise the charcoal burners and control the traceability of their products. Despite this desire to organise, charcoal making is putting pressure on management transfers, as communities can make charcoal with dry wood within the limits of management transfers.

Finally, with the considerable increase in the population, the need for space, whether for housing or for activities, has risen sharply. Added to this is the high mobility of the population with the various resources allocated to this fact. The case of land clearing within the park and certain transfers of management carried out by the coastal communities, particularly those of *Ambolisogno*, bear witness to this.

If we just look at the surface area of the earth and the world's population, we see that the earth's capacity is far from being reached. But the problem is that the earth's resources cannot support this population. In fact, if we assume horizontal construction, i.e. if the world's population continues to grow at the same rate, an increase in population implies a search for new habitable space, and habitable space means development and therefore the transformation and destruction of pre-existing resources.

Demographic pressure is therefore largely responsible for the environmental degradation of the *Mahafale* landscape. It will block the success of the management transfer.

But what part does the economy play in blocking the transfer of management?

The second factor of environmental degradation that will hinder the success of management transfer is the over-exploitation of natural resources due to economic activity. These can be divided into two categories depending on the level of development of the country concerned.

In industrialised countries, for example, much of the environmental degradation is due to consumer behaviour. Excessive consumption leads to the over-exploitation of renewable and exhaustible resources, and above all to the greenhouse effect and the hole in the ozone layer. With around 30% of the world's population, they consume most of the world's environmental resources and pollute a great deal. Around 70% of CO_2 emissions come from industrialised countries [BONTENS H. and ROTILLON G. 1998].

All this pollution is due to the specialisation of northern countries in industrial activity, reflecting what Ricardo called "comparative advantage", i.e. a country specialises in the activity that gives it a comparative advantage. The economies of developed countries are based on industry, and economic growth comes from the exploitation of natural resources and the domination of nature by technology [RAMIARISON H, 2008]. However, most of the technologies used are highly polluting and require a lot of natural resources, such as oil, to operate.

If the destruction of the environment can be explained by increased industrialisation in developed countries, what about undeveloped countries?

In contrast to developed countries, economic activity in poor countries is dominated by agriculture and the export of natural resources such as wood, rice and maize.

Industrialisation in non-developed countries is low, and if they had maintained the same rate of exploitation and the same production methods, there would have been no destruction of the environment and natural resources.

But this was not the case. In their desire to reach the same level of development as the industrialised countries, the poor countries imitated the production methods of the industrialised countries[129] . Poor countries became ever greater consumers of natural resources and ever greater polluters.

In the case of Madagascar, more specifically the *Mahafale* plateau, it has suffered from drought caused by climate change. Climate change is the result of the greenhouse effect. The *Mahafale* community has no alternative to drought other than deforestation for traditional agriculture and the production of charcoal and planks.

Finally, the last factor blocking the transfer of management is the institutional factor. In this section, for ease of understanding, we have chosen the state and the market as institutions. Indeed, it is at the level of these two major types of institution that the major decisions are taken. This section looks at the failure of these institutions in environmental management.

Firstly, for classical economists, the market is a self-regulating mechanism. An "invisible hand" coordinates and guides individual actions towards social harmony [SMITH A. quoted by MANDRARA E. T., 2005]. But the reality is quite different.

In reality, there is no market for the environment, although some natural resources are traded. The first market failure is that individuals underestimate the value of natural resources, leading them to over-exploit them. This fact is reinforced by the underestimation of the costs involved in exploitation, with individuals taking account only of the costs borne by themselves and ignoring the costs borne by others. A car owner, for example, only takes into account his expenditure on fuel and maintenance and ignores the effects of using his car on others, such as noise and air pollution. This attitude is often referred to as externality[130] .

Market failure is also linked to its characteristics. When we talk about the market, we always refer to the law of the market, according to which the price is set on the market by the confrontation of supply and demand. But one thing that is not said is that this market is subject to numerous imperfections, including asymmetric information, uncertainty and risk. All these

[129] H. Ramiarison, 2008. Op. cit,

[130] It refers to an economic situation in which the action of one agent has a positive or negative impact on the situation of another agent not involved in the action, without the latter being fully compensated or having to pay for the damage or benefits caused.

market imperfections lead to natural resources being undervalued and overexploited. Indeed, for example, given these market failures, people may value the present excessively and use natural resources intensively for fear that their present values will be higher than their future values. Another example is the over-exploitation of natural resources when information on these resources is not available (as in the case of Madagascan rosewood, even though we have no statistical figures to confirm this, for more than a decade). Agents may over-exploit resources because they are unaware of their characteristics, such as the rate of growth and renewal. These facts are accentuated by the irreversibility of actions already taken.

Market failure in the management of natural resources can also be explained by the role played by prices. Prices can rise when resources are scarce, stimulating the search for new resources, the use of substitutes and a reduction in the consumption of natural resources. However, this role of price regulation is limited.

Firstly, when there is no efficient substitute, i.e. in terms of cost or ease of use, even if the price increases, this increase is followed by only a slight change in consumption, and this in the short term. In this case, demand is not very elastic in relation to an increase in the price of natural resources, as we can see in the case of oil.

Secondly, when the State intervenes in price setting, as it does in agriculture where it applies guaranteed prices, prices cannot exceed a ceiling price and therefore prices can no longer fulfil their regulatory role.

Finally, this regulatory role of price is no longer valid when there are groups that can influence it either downwards to reduce production costs, or upwards when they want to limit the entry of new competitors, as in the case of a monopoly.

In short, market failure has aggravated the environmental problem. This failure is linked to the imperfections of the market, to its mechanism and also to the behaviour of individuals. But what about the state?

The role of the State is to ensure that the needs of the population are met. In general, to satisfy these needs, given that needs are unlimited while means are scarce, the State undertakes actions that aim to act on these needs and direct them towards a deliberately preferred choice.

In the environmental field, this role mainly consists of encouraging individuals to behave in ways that improve collective well-being, for example by applying laws and redistribution policies.

Firstly, the failure may stem from the absence of an environmental management policy. It is difficult to find a desirable and acceptable policy and to implement it.

The second failure of the State is its inability to match policy to the needs of the population. Demand from the population is constantly changing, and differs according to social categories and their lifestyles, whereas State policy is inert and bears no relation to this demand. However, the effectiveness of State policy depends on its ability to satisfy any demands that may arise, and the legitimacy of the State also depends on this.

The failure of the state to manage natural resources can also result from the antagonism between environmental management policies and state policies on activities that require natural resources. For example, on the one hand, the state has drawn up a forest protection policy and, on the other, it promises an economy based on agriculture or mining. In this case, to encourage the economy, the state can grant subsidies to farmers or guarantee the price of production to encourage producers to increase their output. In order to increase production, farmers have to over-exploit the soil, and if the area available for cultivation is insufficient, they have to look for new land, which leads to slash-and-burn cultivation. This is known as subsidised destruction. The State is therefore forced to make a choice between the well-being of the population or economic growth, and it should be noted that this situation is only valid if farmers are motivated by offers from the State.

In short, environmental degradation and the over-exploitation of natural resources have been caused by exponential population growth, especially in undeveloped countries, by production methods and also by the failings of institutions such as the market and the state.

7.2 Scope and limits linked to the policy pursued by the players

7.2.1 Environmental NGO policy

Following the promulgation of the Gelose law, imposed indirectly by donors, Madagascar no longer neglected ecology. When President Didier *RATSIRAKA* returned to power in 1997, he adjusted his rhetoric by promising a humanist and ecological republic. In July 2002, President Marc *RAVALOMANANA* came to power promising rapid and sustainable development that would reconcile the language favoured by donors with that of urgency. In Durban, the President opted for sustainable development that promotes the conservation of natural resources by increasing the surface area of protected areas in Madagascar.

As a result, over the last ten years or so, there has been a proliferation of foreign and Madagascan NGOs working in the field of environmental protection[131] and agricultural development. The emergence of these intermediary players between the global and local levels is a response both to the demand of international donors for NGO-type partners to implement Madagascar's ambitious environmental policy, and to the mobilisation of urban intellectuals in search of rewarding jobs.

Indirectly, development supports the population, but it also serves NGOs directly. In conjunction with the latter, development basically helps the most advanced economic sector, which is the urban population.

NGOs are generally essential because they are the actors, the prime contractors of much more global projects financed by international donors or the State, etc. i.e. they are the ones who will set up the projects at local level. To this end, there is certainly international competition between NGOs. Each NGO operating in Madagascar has its own operating methods and objectives. These differ according to the requirements of their backers, even if they work on the same issues.

As far as the *Mahafale* landscape is concerned, several environmental NGOs are working to conserve the *Tsimanampesotse* salt lake.

The MNP is the park manager. It is therefore primarily responsible for the park. The amount of funding injected by donors depends on the surface area of the land to be conserved. To this end, MNP managers have done their best to extend the park's area. For example, during the lean season in 2010, the MNP mobilised the *fokonolona* to build a track along the cliff, and then

[131] When we talk about the environment, we're talking about biodiversity

decreed that this track would be the boundary for the extension of the national park, even though the new boundary encroaches on the CoBa conservation areas. The CoBa felt betrayed by the MNP. The extension of the park then led to a reduction in traditional access to the forests.

In 2 years, the CoBa have used up the wood used to build coffins *(mendoravy)* in the conservation area. To obtain *mendoravy*, local people currently have to apply to the MNP for permission to cut the wood.

Specific measures need to be considered regarding access to and sustainable management (by the MNP) of the trees needed to build the coffins. MNP officials have reached an agreement with the *Itampolo* communities on the use of the park's areas to quell any possible uprisings. They named this agreement the Ten Commandments of *Itampolo.* The MNP was then obliged to build a few wells for the coastal communities, even though the quality of the water was poor due to the high salinity of the coastal water. The priority for the communities, however, is to have traditional, free access to the forests to meet their needs.

The WWF, SAGE and GIZ are working with the CoBa. These are supposed to be a belt and a shield that will protect the park through the transfer of management. With the grassroots communities it has supported, the WWF has focused its efforts on literacy and the adoption of a new cultivation technique to improve production. This is the SCV method or cultivation under green cover. In fact, WWF is supplying a number of communities known as pilot farmers who are popularising this new technique.

To train the pilot farmers, WWF convened two members of each community that manages forest resources. WWF will have to change this strategy, i.e. adopt proximity training, as this is more relevant than summoning farmers' representatives to a district capital to attend an action training course.

SAGE, for its part, financed the patrols of the grassroots communities managing the management transfer so that they could achieve their objectives.

As for GIZ, it has financed the seeds and field work activities of members of grassroots communities managing forest resources. As a result, there is a kind of hidden competition between the grassroots communities around the park.

It is therefore necessary to promote better coordination between the various support organisations, in particular to avoid leakage between CoBa.

It should be noted, however, that this phenomenon of flight attests to a certain appropriation of the management transfers by the population and the beginning of a certain dynamic motivating the CoBa to take charge of its own land.

NGOs are not in the habit of reporting to the local forestry administration, despite the latter's status. As far as activity reports are concerned, the NGOs send their[132] to the donors. The local forestry administration must report to the NGOs when the latter have entrusted a mission to the local administration.

The divergent approaches of the technical and financial partners mean that the support organisations' interventions are weak, even if they are working on the same theme and in the

[132] In general, NGOs produce reports on the basis of the requirements and results expected by their donors.

same sector of activity. As a result, the support organisations[133] do not have uniform procedures, despite coordination efforts at regional or national level, with each organisation having its own objectives to achieve. This situation is deplorable insofar as local grassroots communities take advantage of it to obtain support that does not always meet their real needs. This attitude makes it difficult to achieve effective ownership and lasting results.

7.2.2 Forest administration policy

The process of transferring management aims to give local communities responsibility for the sustainable management of natural resources. This process involves the State delegating some of its powers to decentralised local authorities so that they can manage their resources rationally. This is known as empowering grassroots communities at all levels by transferring responsibilities and powers from the State.

To this end, local communities that use natural resources can negotiate a resource management contract with the government, based on a set of rules, to manage these resources for an initial period of three years, renewable for subsequent ten-year periods. It is based on three fundamental principles: voluntariness, demonstrated by a request for transfer from the grassroots communities; subsidiarity, by seeking complementarity between the stakeholders; and non-exclusion, by involving all the different social strata in the area concerned.

This process of transferring the management of natural resources to local grassroots communities takes place within the framework of the Gelose law, which came into force in 1996. This law regulates all the different types of renewable natural resources, such as forests, lakes, pastures, protected areas, etc., but our analysis focuses simply on forest resources, which were the subject of decree 2001-122 on the transfer of forest management, known as contractualised management (GCF). On the other hand, the successful application of this law takes into account the establishment of well-coherent administrative structures.

The management of natural resources is the responsibility of numerous administrative structures, and this practice makes inter- and intra-institutional collaboration difficult because of the difficulties in coordinating activities. Indeed, it is probably the lack of clear procedures that prevents different divisions within the same administrative department from having the same vision of objectives and working with coherent approaches. This situation is made even worse when the departments are geographically distant and do not come under the same ministry.

Formalising policies by promulgating a law or legal text at least makes it possible to provide for the institutional arrangements that should encourage better co-ordination of actions. However, very few policies have been formally formulated, and in any case their dissemination is not always automatic. Despite the various interministerial coordination meetings at national level (council of ministers or interministerial committee), natural resource management poses obvious coordination difficulties, due to the lack of mutual understanding of the objectives and implementation strategies of each of the partners.

These difficulties are already apparent at national level within a team that is supposed to be homogeneous because it is part of the same government and therefore supposed to achieve the

[133] It should be remembered that three support organisations are currently involved in the *Mahafale* Plateau, namely GIZ, MNP and WWF.

objectives. At regional level, they may be exacerbated by the entry into the fray of other players (NGOs, farmers' associations, forestry operators, timber companies) who do not necessarily have the same vision or the same interests in natural resource management.

At local level, the forestry service is stifled by limited resources and means. In the eyes of local stakeholders, the forestry service remains the main representative of the state responsible for forest management. The visibility of the forestry service's support is an important guarantee of the legislation of community management law. However, it does not always have the resources to provide advisory support and local monitoring for the implementation of management plans and to support local grassroots communities at the appropriate time.

Moreover, despite the efforts made, the decentralisation process is still slow. The slowness of this process is hampering the decision-making of the State representative and will upset the trust between the forestry services and the grassroots communities.

In particular, the delay in the official evaluation of the first three probationary years for the renewal of the management contract calls into question the legality of the power of the local grassroots communities. This "unclear" situation is a demotivating factor and is blocking further implementation of the management plan.

We can confirm that the State is not yet ready to transfer the management of these forest resources to local communities. Because of pressure from donors, the central government has hastily decided to transfer the management of forest ecosystems to local communities without putting the necessary mechanisms in place. In the field, the forestry service is often obliged to make corrections or recommendations. Nonetheless, there is some regret on the part of forestry officers, who feel that they have lost their prestige with the people being managed and feel that they have been forgotten in the activities to be undertaken in the field.

7.2.3 The policy of the municipalities and the AICPM

7.2.3.1 Local authority policy

The commune to which the forest is attached is the local authority in whose area the managed forest resources, which are the subject of the management transfer, are located. The commune is in constant contact with the forestry administration and the local community that manages the forest resources. It should therefore be heavily involved in the management transfer process, even if it is not a signatory to the contract. It must arbitrate any disputes arising prior to the contract. In addition, the commune, through the intermediary of the president of the council, is responsible for reconciling the parties in dispute before referring the matter to the competent court or resorting to arbitration.

Elected representatives in decentralised local authorities have powers that enable them to take initiatives to promote the development of their locality.

However, these initiatives are often hampered by the limited resources available in the communes. These resources, consisting mainly of state subsidies, barely cover the running costs of these decentralised authorities. In addition, the level of funds transferred to the communes is very low. The small amount transferred to the communes has a considerable impact on their investment capacity. It reduces the scope for grassroots development projects. Fiscal resources are also low, and insufficient to replenish commune coffers.

As a result, each local authority is trying to resolve this shortage of financial resources and is rushing to work with financial and technical partners. Taking advantage of this situation, the latter have imposed conditions for granting aid and funding. Environmental protection is one of these conditions.

The local authority's elected representatives will then include environmental protection projects or actions linked to environmental protection in their municipal development plan as a wink to donors.

In fact, having a transfer of natural resource management in place makes it easier to obtain funding or enhances the municipality's image in the eyes of technical and financial partners. Thus, commune representatives will encourage local communities to request a transfer of forest resource management in order to obtain aid and funding. Ordinances 73-009 [19 March 1973] and 73-010 [24 March 1973] relating to the powers and responsibilities of the *fokonolona* stipulate the responsibilities of the fokonolona with regard to the governance and management of natural resources located on their land. The initial principle of transferring the management of natural resources was based on assigning the management of these resources to local communities, commonly known as "*fokonolona*". Unfortunately, the voluntary principle included in law 96-025 seems to have minimised the assumption of collective responsibility, resulting in a clash within the community concerned, with some of them registering as VOIs supported by international organisations and KASTI[134] supported by the local forestry administration, which the State considers to be the only legal manager entitled to the benefits derived from natural resources.

In addition, the representatives of each commune compete to be the model for environmental protection, and relations between the local authorities and the population differ from one social group to another. The lack of a structure for dialogue at commune level in which citizens can express themselves freely may be a major cause of this. In the event of problems between communities, there is political recuperation by antagonistic parties (pro-regime and opposition). They compete to control internal conflicts over the management of forest resources.

Despite support from the commune for the creation of associations and the revitalisation of farmers' organisations to manage forest resources, the commune's objective is still to seek collaboration and funding. It does not yet intend to set up a platform that could be a forum for dialogue between the authorities and the local population, because of the lack of resources, but also perhaps because it is not its priority.

As a result, CoBa and KASTI are in competition with each other. In the relationship between the municipal players, there is a separation that does not allow them to concentrate their efforts to act in synergy and together for the integrated development of the municipality, because each is acting independently even though the objectives are the same.

There is no denying the efforts made by local authorities, existing farmers' associations, local economic operators, government bodies and donors to promote development and conserve biodiversity, but the lack of complementarity means that these actions are not sustainable.

[134] The Komitin'ny Ala Sy ny Tontolo Iainana is a group of farmers who conserve biodiversity on their land. It was set up by the forest district chief. It has not signed a management contract either with the forestry administration or with the officials of their commune. It does not have a management tool. In general, the KASTI has not received any training in forest resource management.

7.2.3.2 AICPM policy

As we explained earlier, AICPM is an association of the 13 communes in the *Mahafale* landscape, an environmental association that operates like an NGO. The association's only source of income is membership fees. The members are not proactive in paying their dues. The association therefore waits for help from NGOs working in the environmental field to undertake their activities. It sells its services to organisations working in the *Mahafale* plateau. It is trying to bring together the VOIs that have received funding from *SGP/Tany Meva* so that it can co-manage its funding with the grassroots communities. However, the grassroots communities have not accepted this proposal.

7.2.4 The CoBa policy

Initially, the grassroots communities of the *Mahafale* plateau wanted to protect the forest resources around the *Tsimanampesotse* park in order to benefit from them: either help with their day-to-day problems or funding. In other words, protection of forest resources in exchange for aid or funding. In other words, in relation to the objectives of the technical and financial partners, the grassroots communities have a single objective: to satisfy their basic needs because they may have known that the forest on the plateau is protected by the spirits known as *Tambahoake*, with the *ombiasy* and *mpitankazomanga* keeping watch over them. They believe that whether or not they are managed using the Gelose law, the forests of the *Mahafale* landscape are already managed traditionally thanks to the strong presence of habits and customs. Consequently, their motivation or degree of participation in the activities to be undertaken, whether ordered by the technical and financial partners or included in the management plan, depends on the volume of funding. However, since the implementation of the management transfer in the *Mahafale* plateau, the grassroots communities have not received any funding from international environmental NGOs such as WWF, but they have been provided with seeds.

The transferred forests have not been properly managed. Local people continue to use the forests to meet their needs. They take wood without asking the management committee, and transhumants have come to build cattle pens and dwellings without wanting to respect the *dina*[135] . In fact, the transhumant herders have a right of way through the management transfers to go to the transhumant arrival zones. However, only one very powerful person (the former mayor of a neighbouring commune who owns thousands of oxen) claims the right to graze in a management transfer because he considers it to be an arrival zone.

In short, grassroots communities are still used to freely using the natural resources around them. The implementation of a transfer of management to groups of individuals in the communities is likely to encounter blocking factors. The social exclusion of disadvantaged groups is both a blocking factor and a consequence of the process. This exclusion was implicitly caused by the local social structure, which hardly allowed the expression of dominated social groups such as the women of the *Mahafale*. The support provided was unable to resolve this delicate situation.

In addition, the unresolved internal conflicts generated by the divergence of interests between the local grassroots communities within their union acted as a brake on the development of

[135] The *dina* is not well known by the general public. It has been recommended that it be publicised more widely. It should be remembered that ritualisation normally makes this publicity possible.

community dynamics. In the *Mahafale* plateau, the latent conflict between the *mpiavy* and *tompontany* over the management of forest resources is far from resolved.

Despite this attitude on the part of the communities, many of them would like to conserve the forest resources because they believe that doing so will ensure their rainfall.

They are also convinced that biodiversity only has value if it is given value. They live off natural resources. They depend on natural resources. Management transfers can therefore be a tool for conserving biodiversity, even if in some places they simply displace the problem. Moreover, a single grassroots community cannot fight the future, but the investment must be made over the long term and in a network.

To better manage their resources, the grassroots communities of the *Mahafale* landscape have joined forces with a platform that is more powerful than they are. This platform is called *TAFO MIHAAVO*.

The grassroots structures that have the same problems unite on this platform because, in general, they have deduced that :

- the assumption of responsibility by grassroots communities is weakened by the inadequacy of their powers and resources in the face of those who are more powerful or have more power than they do;
- the laws governing the exploitation and protection of natural mining, forestry, fishing, marine and other resources are not consistent;
- the benefits of natural resources are distributed unfairly ;
- Respecting social disciplines is very difficult because many families depend on natural resources for their livelihood.

To solve these problems, the members of the platform are going to focus all their efforts on setting up and implementing a national strategy with a collaborative relationship and the management of natural resources at grassroots level. The platform's vision extends to 2020 and focuses on a clean environment and the equitable management of natural resources as a lever for sustainable development.

Grassroots communities will be responsible, efficient, autonomous and respected by all in terms of their powers, competences and values. To achieve this, management transfer alone is no longer enough. To eradicate the despotism of some when it comes to natural resources, we need to put in place a structure in which there is "a division of powers and responsibilities between the authorities, grassroots structures", the private sector and civil society, "for better governance". It must also include the power of grassroots communities to lodge complaints against any plunderers of natural resources, who will be punished by competent, independent and fair judicial authorities set up for this purpose.

The grassroots communities will take charge, together with the other stakeholders, of activities to protect the environment, ensure their subsistence and improve their living conditions in general. As a result, they are putting in place the structures and the management plan for their land, and the official status of the VOI as well as the Gélose law and the MECIE ordinance will be revised (Mise en Cohérence des Exploitations avec l'Environnement) so that everyone can use and protect Madagascar's land and natural resources and benefit from them in an equitable manner.

Lastly, the grassroots structures "exchange experiences and communicate transparently" locally, nationally and internationally, so that they can develop rapidly, advise and implement the activities assigned to them immediately.

This network of grassroots communities has an ambitious challenge. It's true that having a positive vision of natural resource management policies is great, but what we're interested in is whether illiterate communities like those on the *Mahafale* plateau can manage to meet these challenges on their own.

By showing the scope and limits of the approach linked to the policy pursued by the players, we can affirm that these players each have their successes and, above all, their limits. Moreover, the Gelose law and the GCF decree that they have implemented also have their limits.

The secret to the success of the transfer of management of the *Mahafale* plateau lies in the communities' respect for customs and traditions, as most of the forests transferred to the grassroots communities include sacred or taboo forests. Our ancestors already managed these forest resources. Their management was based on dissuasion through threats, participatory management and awareness-raising, showing that the forest is ultimately the last resort of the poor.

Thus, the policy of participatory forest management is not new. The way in which it has been implemented has changed from one regime to another. At present, the transfer of natural resource management has taken place thanks to the implementation of the Gelose law and the GCF decree. Donors have entrusted the implementation of the latter to national and/or international non-governmental organisations.

Indeed, the conservation of renewable natural resources is becoming a real business, a real international competition between NGOs. It is not the natural resources that concern them, but the market. In the case of the *Mahafale* Plateau, the intervening NGOs do not have the same procedure despite the coordination efforts undertaken. The divergent approaches of the technical and financial partners mean that the support organisations are weak, even though they are working on the same issues and in the same sector of activity. This situation is deplorable insofar as local grassroots communities take advantage of it to obtain support that does not always meet their real needs. Such an approach makes it difficult to achieve effective ownership and sustainable results.

CONCLUSION ON THE SECOND HYPOTHESIS

To repeat our second hypothesis: "Traditional social organisation and customs ensure the stability and success of management transfers in the *Mahafale* plateau".

Our analysis seems to confirm this initial hypothesis. The central government does not seem to be ready to hand over the management of natural resources to local communities. This shortcoming has had an impact on decentralised local authorities. Thus, transferring the management of renewable natural resources remains a fad for the communes, as their objective is still to seek collaboration and funding.

What's more, from the outset, the grassroots communities managing the natural resource management transfers on the *Mahafale* plateau have a dual objective when it comes to protecting natural resources: to be a role model for other communities and also to benefit from funding to meet their basic needs.

In addition, management transfer support organisations are obviously the prime contractors for renewable natural resource conservation projects. In other words, they are the ones who will set up the projects at local level. However, these are the people who are most sensitive to the realities they encounter in the field, not the *mpitan-kazomanga* and *ombiasa.* The latter are excluded from the management of these forest resources, even though they are the real custodians.

In the *Mahafale* plateau, the last stage in the management transfer process, which is ritualisation[136] , is omitted by the support organisations, with the group of people forming the management committee becoming the new manager of the ancestral forests *(alan-draza).*

As a result, the actions of CoBa and KASTI are inconsistent and often even antagonistic. In the relationship between local players, there is a separation that prevents them from concentrating their efforts to act in synergy and together, for the integrated development of their municipality, because each is acting independently even though the objectives are the same. That is, to manage forest resources in return for funding.

All those involved in this management transfer are simply confiscating traditional management by the *ombiasy* and *mpitankazomanga.* The success of this management transfer has long been based on respect for the people of *Tambahoaka* (a guardian spirit of the forests of the *Mahafale* plateau) and respect for customs and traditions.

Local people continue to use the forests to meet their needs. They take wood without asking the management committee. Transhumant herders have come to build cattle pens and dwellings without wanting to respect the *dina.* Trafficking in rosewood in areas managed by the CoBa is intensifying. The transferred forests have not been properly managed. The rest of the forests on the *Mahafale* plateau are taboo forests or sacred forests whose use requires the sacrifice of zebus.

[136] Ask for a blessing and a free pass from the ancestors and forest spirits *(Tambahoake)*, as the CoBa will take over the management of forest resources through management transfer.

Chapter Eight: Sustainable development and conservation of natural resources

All the actions undertaken by the various economic agents are aimed in one way or another at improving the socio-economic and political situation, which is none other than development. There are many different concepts of development, ranging from simple economic growth through social development to sustainable development.

In this chapter, for further clarification, we will look at whether we can achieve sustainable development without or with conservation of natural resources and the ideas associated with it.

8.1 Sustainable development

Before going into detail, we need to clarify the difference between "*sustainable development*" and "*sustainable development*".

In general, development should be sustainable, i.e. last for a long time. But to be sustainable, it has to be sustainable.

Here, sustainable development is development that can be supported by man, nature or natural resources. We can therefore say that sustainable development is synonymous with "*ecologically viable development*"[137] .

Sustainable development is defined as "*development that meets the needs of the present without compromising the ability of future generations to meet their own needs*"[138] . This concept has been developed since 1980, when it was translated from English as "*sustainable development*" in the work "*The World Conservation Strategy*" by the International Union for Conservation of Nature (IUCN). Its aim is to reconcile ecological, economic and social considerations and to find an optimum balance between these three elements, so that overall productivity offsets the losses caused by economic activities that deplete natural resources.

Development has had various definitions and names. The current term is sustainable development. In what follows, we will look at the elements taken into consideration in this new type of development.

8.1.1 The systems approach to development

A system is made up of several elements that are closely interdependent and organised in a certain way. For Condillac, a system is an arrangement of elements in a mutually supportive order; the whole is in each element, and each element has no meaning unless it belongs to a whole. It is in this sense that the new concept of development should be understood. As a fight against poverty, development must take into account three important elements: the economic, the social and a new element to which the two previous elements are linked: ecology.

8.1.2 The ecological aspect of development

Aware of the destruction and degradation of ecosystems (desertification, global warming, etc.) and the associated negative externalities, the need for better management of natural resources

---

[137] World Bank, 1992, World Development Report.

[138] G. Grarnier and Y. Veyret, 2006. *Développement Durable: Quels enjeux géographiques?* La documentation Française.

has become a priority. Development must take the environment into account, because it is not enough just to satisfy the needs of the present but also those of future generations, and it must respect the environment. We can do this by answering the question "*What should we leave to our children and grandchildren so that they have the maximum chance of not living worse than we do?*

The environment must therefore be preserved, and actions must be taken to ensure the sustainability of natural resources when seeking economic growth, so that future generations can benefit from it.

8.1.3 The economic aspect of development

Economic growth is a necessary condition for development. We need to encourage economic growth, an increase in production to satisfy the needs expressed by the population. This economic growth must take account of two other factors: it must result in an improvement in the population's standard of living without harming the environment. In the past, economic growth consisted solely of increasing production while ignoring its impact on the population and, above all, on the environment. Today, however, it must assess the externalities that the population and the environment may suffer.

We might think that you can't have economic growth and environmental conservation at the same time, because economic growth aims to increase production by exploiting natural resources. So we have to choose between economic growth and the conservation of natural resources. However, this is not always the case. In fact, economic growth and environmental conservation are not contradictory for developed countries. For economic growth to be ecologically viable, we need to produce differently; we need to change the way we consume natural resources and use production technologies that are more respectful of the environment but still deliver the same desired growth.

However, the implementation of the natural resource conservation policy will lead to a slowdown in economic growth. This slowdown in economic growth can be explained by the fact that there is no automatic correlation between economic growth and a change in production methods. The slowdown in economic growth continues until it reaches a minimum level corresponding to an effort to conserve the environment.

Before applying the environmental conservation policy, the Madagascan government does not take into account the economic development of everyone. There are no measures to support this policy. What's more, they have turned the agricultural high schools into a programme (run by non-governmental organisations). This is what we call a strategic error.

If this is the economic aspect of development, why is the social aspect important?

8.1.4 The social aspect of development

Some people say that environmental management is important for development. Without good environmental management, we cannot have development. However, the reverse relationship is also important, and this is perhaps where the emphasis should be placed, because without development there can be no good environmental management.

The majority of the world's population, especially in developing countries like Madagascar, is poor. Poor people do everything possible to satisfy their essential or basic needs. They survive by exploiting natural resources. They make very little use of modern technology. So if they want to increase their production, they have to exploit natural resources to the maximum, including forestry for consumption.

If we are to have any hope of managing these natural resources, we need to lift people out of poverty, give them a better standard of living and get them to stop depending on natural resources. However, this seems impossible because the population is constantly growing and demand is increasing. As a result, nature continues to be exploited. For us, the fight against poverty is essential if we are to conserve the environment.

To conclude, if we want to talk about sustainable development, we need to take into account the following three pillars: ecology, social issues and the economy.

For development to be sustainable, it must be socially and ecologically liveable, economically and socially equitable, and economically and ecologically viable. We therefore need to strike a balance between these three pillars, without which development cannot be sustainable.

But from this relationship a problem arises: that of equity between the population, between countries and above all between generations. The solution lies in the optimal use of natural resources.

Sustainable development has now become a global objective. To achieve it, we need to solve the problem of equity. Equity arises as soon as we talk about sustainable development, because it implies a balance between the following three elements: economy-society-environment.

8.2 Does sustainable development mean eternal impoverishment for Madagascar?

8.2.1 Evolution of the concept of sustainable development

The concept of sustainable development has been developed since the first international conference on the human environment in Stockholm (under the aegis of the United Nations) in 1972. The conclusion drawn was to propose a model of economic development compatible with social equity and ecological prudence. This model was named the "ecodevelopment" model. Then, in 1987, Norwegian Prime Minister Brundtland's report "Our Common Future" redefined sustainable development as development that meets the needs of the present without compromising the ability of future generations to meet their own needs.

In 1992, at the Rio Conference, sustainable development was defined as a change in production methods, a change in consumption practices and, above all, the adoption by citizens and industry of everyday behaviour that preserves the quality and diversity of the living environment, resources and the environment. The development model of Western societies is no longer considered to be the only compulsory development model (at least in theory). The following conclusion has been drawn: *"a diversity of situations and cultures must be matched by a diversity of forms of development"*.

After these key dates, the notion of sustainable development was discussed at a number of events[139] . The definition of this concept is no longer on the agenda, but rather the solutions to be presented to avoid possible disasters and preserve the environment.

8.2.2 Appropriation of sustainable development ideas

[139] Through international conferences and symposia

We will be talking about the mobilisation and appropriation of ideas developed at international level on sustainable development in national policy-making processes.

In Madagascar, the adoption of this sustainable development framework was early. As early as 1984, the country was a precursor of sustainable development, initially approached from an environmental perspective. However, sustainability was not taken into account until the early 1990s, with the adoption of an environmental charter and the launch of the Environmental Action Plan (PAE) supported by a consortium of donors (World Bank, USAID, GTZ, etc.). This reappropriation of the sustainable development frame of reference in national policies currently seems to be characterised by a desire to break with sectoral practices of public action, focused on the environment.

This integration can lead to the introduction of new policies that seek to integrate the various dimensions of sustainability, but also to changes in pre-existing policies.

The first programme (1991-1996) was characterised by a centralised approach to the management of the environment and natural resources: zoning of protected areas based on scientific criteria, without any consultation with local stakeholders, and stigmatisation of agriculture as the main source of degradation. Sustainable development then loses its substance as rural development takes second place.

The adoption of these internationally inspired public policies means that the country's leaders will be dependent on and assist the international community, because Madagascar will be dependent on development aid. Rich countries are racing to borrow Madagascar on their terms. This raises the question of the degree of freedom the country has in defining its development choices and public action. The divergence of proposals between scientists and communities is at the root of a deterioration in relations between operators and the local population. Hence Chaboud's comment that *the sustainability and social acceptability of these guidelines decreed at international level give cause for thought* [Chaboud. 2007].

So, have assistance relationships really evolved into partnerships that are supposed to be more constructive? We need to question the reality of current conservation projects.

Chapter Nine: Transfer of forest resource management and its future

The aim of Agar was to restore the rights of nature and the rights over the environment to the population as a whole. In other words, it aims to restore a central place to local populations. So, for some people, it finally makes room for the poorest, but for others who are not in the minority, it is welcome because it will provide a framework for those who benefit from the resources. However, according to the theory of integrated conservationthe objective of the CoBa would be to conserve forest resources and exploit them at the same time. However, in the practice of "development" or "conservation of the forest ecosystem", there is a clear distinction between support projects that aim to conserve nature and those that aim to provide a framework for the productive and commercial use of resources by villagers. The dissociation of objectives, which reflects the division of labour between those involved, is more marked here because the resources concerned are located in the peripheral zone of a national park. In Madagascar, for example, the status of NGOs that can make a profit has meant that agar is sometimes misappropriated to help certain individuals to make a profit for themselves. These individuals are actively involved in implementing conservation projects. Even if they are actively involved in conservation, can they achieve the ultimate goal of management transfer, which is the long-term management of these resources so that their offspring can benefit?

In this last chapter, we will try to verify that *the management transfers studied do not achieve the objectives of sustainable management because of the reduction in the Mahafale communities' use rights, the geographical situation and climate change, which weigh heavily on their land.*

9.1 The future of the forest ecosystem on the Mahafale plateau

The *Mahafale* region is a sub-arid zone with one of the highest poverty rates in the country. Ecologically, it is covered by xerophytic vegetation that is unique in the world. However, because of their extreme poverty, the people living near these forest ecosystems are often forced to exploit and commercialise existing resources, as this is their only source of cash income and is often a prerequisite for survival. They clear land for maize production and make charcoal, which is used in towns as domestic fuel.

What's more, poverty erases human dignity and a sense of civic responsibility, and the poor, as mere do-gooders with no initiative whatsoever, can expect neither satisfactory social development or position, nor any improvement in their material and living conditions.

Poverty, whether intellectual, cultural or economic, creates a feeling of frustration or dissatisfaction and can easily lead to a lack of interest in community responsibilities.

As a result, most of the population feels marginalised. What's more, they're poor. They don't have much responsibility. Female poverty and the problem of youth unemployment (the rush to the sapphire fields and the rice fields of Morondava in search of a better life) accentuate this phenomenon.

The initiative or other forms of involvement of citizens in community action or specifically in action to conserve forest resources depend on their need to develop as people, and on their individual capacity, which is why we will deal one by one with the factors that could determine the future of the forest ecosystem of the *Mahafale* plateau.

9.1.1 The level of education of CoBa members

Most of the communities on the *Mahafale* plateau are illiterate, and this situation hinders the development of oral culture towards written culture. It also limits the ability to negotiate. However, managing a "community association" such as CoBa requires a minimum of organisational, management and even "written culture" skills. In short, the CoBa manager must master the culture of leadership. Ignorance of the latter is a major constraint in the life of an association. It leads to increased communication problems between the leader and the members of the association. In turn, the communication problem leads to a low level of sharing of internal responsibilities between members.

In rural areas such as the *Mahafale* plateau, associations working for environmental protection or other development objectives are experiencing an accumulation of responsibilities committed by those in power. This accumulation of responsibilities by a few leaders is a direct consequence of the lack of human resources.

For example, when support organisations invite people to attend training courses, it is the same individuals who benefit from them, because some CoBa leaders have been seen to act in bad faith. This attitude can be explained by the current existence, thanks to this association called CoBa, of a new generation of farmer leaders who are in the process of being born. However, the break with the old order is not blatant.

The new leaders most often come from rich and powerful families who have been able to finance their sons' education. This is more common on the coast than inland on the *Mahafale* plateau. Nevertheless, some of them are descendants of the *Hova*, who also relied on schools and the Church.

What we find surprising is that the *Mahafale* plateau has been frequented by NGOs for over twenty years, even though the community as a whole is suspicious of them.

Adult literacy and schooling for the children of the plateau so that they can change mentally are not included in the activity plans of these NGOs.

We have also noticed that when young people from the mining sites, even if they are illiterate, return to their villages, their "mentality" changes much more quickly in the mining sites because they are open to modernity, especially those who use the RN7. This state of affairs is easy to understand. Their cultures mix with those of others and they are influenced by modernity ("a business culture").

Young people, who are considered to be the potential and the future of the *Mahafale* plateau, are not sufficiently integrated to be a real part of the working population or to be leaders in their locality. They are often excluded from major development projects, particularly the natural resources conservation project. Their lack of education, due to the high cost of tuition, partly explains their passivity, or rather their lack of initiative. Their parents also prevent them from taking the plunge and taking their future into their own hands, hence their marginalisation.

We can reiterate that development in the context of globalisation requires the adoption of another culture. It is Western culture that could eventually give a chance to the competent regardless of gender and age.

Malagasy society, particularly in the countryside, would then change its itinerary, its trajectory, to have a new vision and a new perspective within the framework of communal development. This requires community unity and integrity. Young people should prepare themselves and convince their elders to work together to manage the economic and environmental situation of the *Mahafale* plateau.

The form of participation in the conservation or management of the forest ecosystem is different from that of community work such as labour-based training. The management or conservation of renewable natural resources requires the participation of everyone.

In addition, Madagascar's capacity shortage index is 0.988[140], according to the 1996 Human Development Report. The female participation indicator is 1.346[141]. This shows that a large proportion of the population lacks capacity and that women's participation in development is very low. The population of the *Mahafale* plateau is not immune to this situation. Intellectual poverty is a barrier to human participation: people cannot take part in the political, social or economic life of their society because they have intellectual limitations, given their level of education.

In conclusion, the duration of funding for a project to conserve and/or manage forest resources around *Tsimanampesotse* Park is a maximum of five years. In the *Mahafale* plateau, the introduction of management transfers to surround the park has already received two grants. To perpetuate the forest ecosystem management activities maintained by the CoBa, the strategies and skills required and the management tools and techniques should be given to the grassroots communities before they are left to manage the transferred resources themselves. However, the CoBa's level of education is still low. Their capacity for good governance of renewable natural resources after phase I of funding for the creation of management transfers around the park was called into question during the evaluation [in 2010] carried out by the forestry administration. In fact, the objective of the donors is to protect the park so that human pressure decreases.

Because of their level of education, the CoBa become wait-and-see. They are always waiting for instructions from the promoting NGOs before undertaking their activities in the management and/or conservation of the forest ecosystem. Otherwise, they do nothing, especially after the support projects have left. They take little interest in patrolling the transferred forests. "*Between phase I and phase II (the renewal and implementation of the new management transfer contracts) of the funding, the rate of clearing increased, both on the coast and inland, even inside the park*", admitted the President of the *Mandrosoa* CoBa, *Tongainoro,* when we spoke to him. The CoBa president's statement can be verified with the help of map 9, a map of land clearing in the park, which we will present in the next sub-chapter.

9.1.2 The cost of patrols and/or monitoring and evaluation

On the *Mahafale* plateau, the forest is the domain of the profusion and diversity of life: it is a resource given to mankind by God. It is a temple, a place for dialogue with the Creator God,

[140] This figure is the projection of 0.353 in 1993. It is multiplied by 2.8 to give an estimate for 2016.
[141] A projection of 0.481. This is multiplied by 2.8 to arrive at this estimate for 2016.

the spirits of nature *(Tambahoake)* and the spirits of the ancestors. These religious concepts feed into the traditional uses of the forest: charms and remedies made from plants, zebu herding, gathering and fishing [Moreau S, 2002].

Patrolling the transferred forests requires appropriate financial resources as well as good organisation (patrol schedule, number and group of men assigned to the patrol), given that for the *Mahafale* the forest is a place of secrecy, where outsiders are not welcome. Most of the *"mpiavy"* (foreigners) who are members of the CoBa will not patrol with the "*tompontany"* (landowners) and the *"zanatany*[142] *"* (children of the land). If they are obliged to patrol with the *tompontany,* the tompontany will not take them to the strictly secret areas inside the transferred forest.

In addition, most of the forests transferred to the communities include sacred forests (reservoirs of charms and remedies, or ancient tombs).

The smallest area of management transfers is 1052 Ha (TG *Soa Ho Anay Andremba, Maroarivo*) while the largest is 42,999 Ha (TG *Magnasoa Tane, Behombe*) and the distance between the transferred forest and the nearest village is 15 km. When the CoBa organise a patrol in the transferred forest, it lasts a minimum of 4 days. The patrol team is obliged to bring provisions with them as they will be camping there. However, most CoBa do not have a budget allocated to this patrol. Those who go on patrol have to pay for provisions.

For example, in its annual work plan, the *Soa Ho Anay* CoBa in *Andremba* must carry out a patrol every 15 days, i.e. twice a month. The patrol team must spend a minimum of 40,000 *ariary* per patrol (cart hire + provisions). The CoBa will therefore spend 80,000 *ariary* per month to carry out this patrol, whereas the monthly membership fee does not exceed 20,000 *ariary* (at *a* rate of 200 *ariary* per person per month). As a result, CoBa members are not motivated to patrol because of this lack of money. Hence the proliferation of land clearing and trafficking in radiated tortoises in the plateau.

The cost per CoBa for a patrol depends on the area of forest transferred and its distance from the village.

The following image shows the proliferation of clearings in the *Tsimanampesotse* park.

[142] The *Zanatany* are former *mpiavy* accepted by the *tompontany* as co-owners because of their seniority in the village. In the old *Mahafale* organisation, the *Zanatany* occupied the south-eastern part of the village in general.

Map 9: Clearance map for the *Tsimanampesotse* park

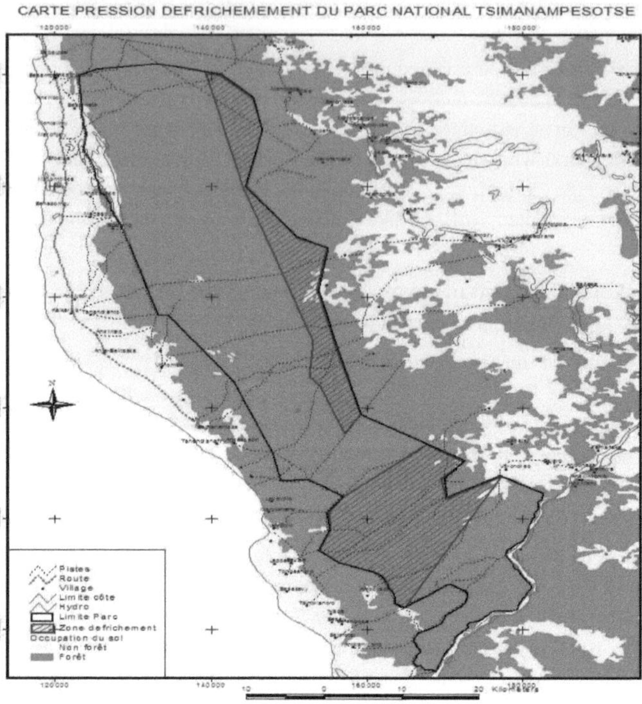

Source: MNP Toliara, 2013

The stripes on this map show the clearing that the park has undergone. On the coast, the clearing in the area of *Ambolisogno*, on the side of the CoBa *Mizakamasy* de *Nisoa* and that of *Mandrosoa, Tongainoro* is the most reported. However, the eastern part of the park has been ravaged by forest clearance. This clearing phenomenon can be explained as follows:

- Firstly, after the park was extended in 2010, the communities voiced their discontent, as most of them felt abused by park officials. It was not only the communities living to the east of the park who cleared the land, but they were also helped by the transhumants from the coast. The latter are voicing their discontent because when they return home, there are no longer any grazing areas for their cattle. As a result, the communities are clearing the former conservation area or the former hard core of the forest transferred to the CoBa;
- secondly, the period in which the clearing occurred was the CoBa management contract renewal period in 2011. During this time, CoBa members, including the board, let go. They no longer felt responsible for managing these transferred natural resources until a new contract was signed;

- finally, since the park has been cleared, this implies that the management transfers, which are nothing other than the park's protective belts, are no longer safe from deforestation. The clearers begin their actions from inside the management transfers towards the outside, i.e. they attack the former hard cores of the CoBa. After that, they ravage most of the area of the management transfer and leave the rest, so that outside the forest there is an intact forest. The CoBa and CLP[143] , together with park officers, do not patrol these areas and the clearers take advantage of this situation.

It is for this reason that we argue that the management transfers studied do not achieve the objectives of sustainable management. Moreover, the CoBa will experience, or are currently experiencing, the end of the project to support the CoBa in the conservation and management of forest resources. There will no longer be project agents to raise awareness among community members about what they should be doing. *"Poachers and land clearers are not afraid of the CoBa, whatever their status, they even dare to challenge us; so you ask what our future will be, we the CoBa, after the departure of the support organisations, I think we will be as weak as we were in 2009-2011; we can do nothing about land clearing and poaching because we are afraid of them. The assessment after 10 years will be catastrophic if the forestry administration doesn't visit us at least once a month"* [Interview with the President of the CoBa *Maevasoa, Zamasy*].

In addition, evaluating management transfers is not at all an easy task.

Firstly, like the patrol, it requires appropriate financial resources, the mobilisation of which depends on the implementation of an appropriate financial policy and the genuine involvement of support organisations and donors.

Technically, establishing time zero (t_o) creates a problem and will also compromise the evaluation of management transfers, as these show very disparate trends over time.

Firstly, however, there is a general problem arising from the very nature of monitoring-evaluation systems. Like any information system, they are designed to meet the specific expectations of certain stakeholders. In the case of the Gelose programme or the GCF decree, there are many stakeholders and their interests are often divergent.

The wide range of issues involved in management transfers makes it difficult to develop a monitoring and evaluation system that can be invested in by all the stakeholders. These difficulties are reflected in the identification of impact indicators, and therefore in the identification of shared objectives and the ways in which they are to be achieved.

Be that as it may, the objectives assigned to management transfers under the Gélose law and the Madagascan environment charter remain "the conservation of biodiversity" and "the improvement of living conditions for the population".

In addition to impact indicators, particular attention should be paid to activity indicators. They can be used to highlight process malfunctions and, as in the case of the management transfer information system, to engage stakeholders in a "quality approach" to gradually improve governance.

---

[143] CLP is a local park committee. This committee is recruited from the CoBa office member in their locality to help the park officers patrol the park through the transferred forests.

Similarly, compliance indicators must be compiled. This enables management transfers to be characterised, particularly with regard to the compliance of all actions carried out in relation to legal and regulatory provisions, including relations with the forestry administration.

Secondly, it must be admitted that it is not easy to evaluate a system of governance. It involves taking into account a large number of interactions between scientists, support organisations, administrations and citizens, and these interactions constitute a complex system. Although these interactions are not always generalised in the context of management transfers, and are partial and/or scattered, they do exist. This type of analysis has great potential, both for research and for support organisations and the forestry administration, because the objective evaluation of management transfers remains a concern for all stakeholders in the implementation of forestry policy.

In fact, the mobilisation of such research can only tend towards interdisciplinarity to tackle a subject as complex as that of the impact and monitoring of management transfers.

By way of illustration, the important roles of geography and remote sensing in relation to spatial analysis complement the economic, ecological, social and institutional approach. However, these spatial indicators could still be better formalised.

Much remains to be done in this area. Spatial information needs to be processed as part of a genuine information system based on an interdisciplinary approach from the outset (i.e. when classifications are being drawn up). This brings us back to the questions: what are we really trying to observe, monitor and evaluate?

The various reports on the *Mahafale* plateau that we have deciphered have highlighted the opportunities and possibilities for monitoring many things. However, the discussions held upstream of the research on the identification of the various objects that can be characterised, monitored and evaluated show the need for the involvement of researchers, practitioners, operators, local communities and decision-makers, right from the design phase of the monitoring and evaluation system.

In all cases, it is important to link all the different approaches to legal issues, including those relating to land tenure, which are closely linked to the duration of management transfer contracts. At present, Law 2005-019 on land tenure reform, in particular Article 38[144] , does not define the status of land subject to a management transfer contract, nor that of land subject to the forestry regime, nor even that of land in protected areas, since these three categories are extremely intertwined. In order to have a coherent vision for the future, it is therefore urgent to invest in these specific regimes, which are necessarily in harmony given the interactions between them.

All these considerations show the amount of work that remains to be done before reliable, indisputable, understandable and easy-to-establish indicators can be developed and provided.

---

[144] In implementing the provisions of this law, account must be taken of areas that are excluded because they are subject to special legal provisions. These areas include :

- land in areas reserved for investment projects;
- land falling within the scope of legislation relating to protected areas;
- land used to implement agreements signed under legislation on the management of natural resources;
- land that is legally defined as falling within the scope of forestry law;
- land that is protected under an international convention ratified by the Republic of Madagascar.

These indicators will help to establish the basis for concrete actions to improve the governance of management transfers, but will also make it possible to assess the sustainability of the management of these natural resources.

There is great potential for understanding management transfers as a conservation and development structure, at local, regional and national levels.

Thirdly, all monitoring and evaluation systems need to be harmonised under the leadership of the Forest Administration. In this sense, whatever the avenues mentioned, the Forest Administration is the central player in the implementation of forest policy, and as such remains legally responsible for monitoring and evaluating management transfers. Its assessment is always necessary for the renewal of management transfer contracts. The lack of evaluation of management transfers by the Forest Administration is extremely detrimental: only 192 out of 896 contracts were renewed in the 13 Regions studied [Lohanivo A. C., 2012].

This raises a number of questions about the sustainability of the initiatives undertaken, despite the fact that major investments have been made and that social and institutional expectations are particularly high.

By way of conclusion, identifying the impact indicators of a monitoring-evaluation system makes the monitoring-evaluation activities of support organisations very difficult because, in general, it has to be reiterated that monitoring-evaluation systems are designed to satisfy the particular expectations of certain stakeholders (particularly donors). In the context of Agar and the GCF decree, the stakeholders are not the same as in other projects, because for Agar and the GCF decree, they are numerous and their interests are often different as a result.

In addition, the different "time steps" that need to be taken into consideration mean that the impact of management transfers needs to be monitored and evaluated over time, which is not possible with a project-based approach. The need to approach long "time steps" can only be achieved by relying on permanent administrations, and therefore, a priori, on State administrations. The latter are acknowledged to have insufficient resources.

9.1.3 Local management difficult to get to grips with

The dry forests of the *Mahafale* plateau contain a biodiversity of worldwide value due to their local endemicity. In general, before gelosis and the GCF decree, the names given to the forests were *alan-draza* (ancestors' forest) and *ala faly* (taboo forest) for the sacred forests. This implies that these forests have been handed down by the ancestors from ancient times, which the community is unable to define exactly. It was the ancestors who decreed that part of each forest was sacred or taboo. Sacred forests are burial places that cannot be desecrated, and taboo forests are places that must be preserved because they contain a wealth of fauna and flora. In general, the taboo forest is guarded by the spirit known as *Tambahoake* among the *Mahafale*. The riches of the sacred forests and preserved forests are used as a reserve in times of crisis: illness (medicinal plants), famine (wild yams, hedgehogs, etc.), disaster (timber for important community sites) and for building coffins for the dead.

These sacred forests are managed de facto by the communities. The boundary of the sacred and/or taboo forests is known to the communities because the trees are taller and denser than elsewhere, which demonstrates the effectiveness of biodiversity conservation. Community rules prohibit the use of sacred forest resources for daily needs. They therefore use resources outside the sacred and/or taboo forests.

However, in the concept of secure local management (Gélose), community conservation areas extend beyond resources (forests, lakes, etc.) and include part of the village land governed by the communities. Community members do not want to understand why their sites are eligible for conservation.

Their sites would only be eligible as community conservation areas if the value of their biodiversity, associated with their culture, is significant and they have the will to manage or govern them effectively for their conservation.

They note that, despite its founding principles, local management as implemented in the *Mahafale* plateau has hardly given farmers any decision-making power, nor has it enabled them to defend their interests and meet their needs as before. The management contracts reiterate, at local level, the forestry law in force, hammering home the prohibitions or limitations on clearing, bush fires and harvesting, without proposing any development likely to protect biodiversity by taking account of the ongoing dynamics of anthropisation. These rules are not always judicious.

It should be remembered that the management of natural resources under the current Gélose or GCF is carried out by a tiny group of people known as the management committee, whereas before it was the whole community.

The management of natural resources within the framework of the Gelose or GCF decree penalises the poorest people, who, at least during the lean periods, live off forest harvesting. The ban on slash-and-burn farming, which used to provide a wide range of food and cash crops, is making everyone's situation more precarious. The division of management territories in an area with blurred contours, prey to mobile activities (extensive livestock farming, gathering, etc.), is far from obvious: carried out in a hurry, it does not always respect the old divisions or disrupts the solidarity between allies who allowed each other to use their respective territories. The internal development plans were also out of step with the realities of the farmers.

Those who are used to having free access to natural resources will react to the prohibitions and restrictions. During our visit to *Behomby* (the headquarters of a CoBa to the east of the park), a father who could no longer stand the rules established at the time of the management transfer argued with the management committee. "*...Most of the people here have not accepted that our ancestor's forest should be managed by just a few heads of the village. The forest is not a legacy from your father where you have imposed rules as you please. It's unfair that we can't get the amount of wood we need. If you want harmony in this village, let's manage the forest according to the rules laid down by our ancestors...*".

This excerpt from a father's dispute with the CoBa management committee is just one example of the disagreements voiced by members of the communities of the *Mahafale* plateau. The communities approve of the protection of the forest, but totally reject the ban on logging in the forest. They accept the ban on land clearance, which they are used to, but not the limits on harvesting, which is vital for some families. Even though the contracts give the managing communities the right to exploit the forest resources on their behalf, very few of them are able to do so. The CoBa have projects (ecotourism, timber or corundum exploitation, tree nurseries in the forests, etc.) that they have difficulty implementing due to a lack of technical knowledge, equipment and resources.

Farmers have a dual view of local management: they see it as an exogenous, restrictive or even repressive structure, but they are also sensitive to the fact that it represents a means of strengthening their control over the forest. In their view, measures should be taken first by the CoBa and then by the supporting NGOs on the ground. This dispute reflects people's lack of ownership of the management transfer.

For the time being, therefore, the management of these transfers remains an imposed process, which peasant society is trying to appropriate. It will not be sustainable because natural resources are under constant threat.

The threats to the sustainability of community conservation in the *Mahafale* landscape relate firstly to food insecurity, which has encouraged abusive exploitation, and secondly to the insecurity of the communities in their current management methods. This food insecurity is exacerbated by the vagaries of the climate and the locust invasions that plague the area every year. To solve this problem, the inhabitants along the RN10 are turning to charcoal production, as the town's need for charcoal has increased. What's more, the *Mahafale* communities have no alternative source of income, as the support projects are not doing enough to promote alternative income-generating activities that respect the environment.

Secondly, population growth will have an impact on the sustainability of the conservation of renewable natural resources. We do not have up-to-date statistical data on the demography of the *Mahafale* plateau, but the population is estimated at 613,317, 6[145] up from 219,042 in 2010. In the space of 5 years, the population of the *Mahafale* landscape has increased by 394,272. Demographic growth is a recent phenomenon, but it has an immediate impact on the dynamics of natural resource management. It has led to the saturation of low-lying valleys and *baiboho* areas that can easily be developed, but also to the triggering of major migratory flows to forest areas with low levels of human settlement. It is also forcing the conquest of new land for development at the expense of the forest.

In short, it is indisputable that the needs of these inhabitants have increased following this demographic growth, and that the pressure of human activity on natural resources has therefore increased.

Thirdly, as its name suggests, the *Mahafale* limestone plateau is made up of limestone rocks. Arable soils cover no more than 15%[146] of the plateau's total surface area. Most of this is located within the park and within the management transfers. However, the government, with the support of donors, has decided to protect these areas. It has protected 27% of the entire *Mahafale* landscape. This decision by the State is destabilising the socio-economic development of the communities, most of which are poor.

The fourth point that weakens the sustainability of natural resource conservation is the current shortage of wood species used daily and locally outside protected areas. Communities are therefore obliged to look for them within the management transfers or the park. In addition, the management committees find it difficult to apply sanctions in the form of fines payable to the CoBa or *dina*. They do not dare apply them in the village because they find them compromising because of the family relationship.

---

[145] This figure is based on a statistical estimate of 219042 x 2.8

[146] In Testimonies from Madagascar: "Climate change and rural lifestyles", WWF 2010

Outside the Park, transhumant herders and their animals have suffered from a lack of water resources and meagre pastures. To make up for this, they have to move into the conservation zone of the management transfer inside the park. They will stay there for 8 months, clearing land to grow crops and support themselves.

Some pressures were neglected or even underestimated two decades ago. Today, they represent imminent threats to biodiversity. This is the case with the development of pioneer slash-and-burn agriculture in remote and sparsely occupied areas (including the interior of the park and the management transfer conservation zone) and also with the production of planks and charcoal in the vicinity of the national roads leading to the major towns.

Finally, the last point concerns the key elements of community forest management.

The following factors are the most important if community management is to be effective and sustainable: the creation of and membership in the CoBa must be voluntary; the boundaries of the community forest must be well defined and accepted by everyone; the community management structure must have legal status; there must be a legal mechanism(s) for transferring management rights; the transfer procedures/processes must be simple and easy to understand; the community must have exclusive rights over the resources to be managed; community institutions must be based on the principles of good governance; forest management must be based on the principles of sustainability; the community must have a guarantee of medium- to long-term management rights.

CONCLUSION OF THE THIRD HYPOTHESIS

To repeat our last hypothesis: "management transfers in the *Mahafale* plateau do not achieve the objectives of sustainable management because of the reduction in the communities' use rights, the geographical situation and climate change, which weigh heavily on their land". We can deduce from this that, in the *Mahafale* plateau, most of these factors that could weaken sustainable management are not considered. By way of illustration, the *Ampasindava* CoBa (charcoal producer) and the *Andremba* CoBa (forest resource conservation) are in dispute because the boundaries of their forests overlap. As a result, members of each CoBa are clearing each other's forests to accuse each other.

Furthermore, sustainable conservation is inseparable from consideration of sustainable livelihoods for communities. Accompanying measures or alternatives to conservation are not being put in place and/or implemented for the benefit of the communities on the *Mahafale* plateau. Also, given that the threats mentioned above still persist and that the management transfer support projects that monitored and supervised the CoBa have now come to an end, the increase in land clearing and clandestine trafficking in natural resources will continue as before the evaluation and renewal of the contract signed in 2011. Fauna and flora endemic to the *Mahafale* landscape will be at risk until the new funding allocated to management transfers around *Tsimanampesotse* Park returns.

9.1.4 The proliferation of mining

The expansion of the current mining sector is akin to the beginnings of industrialisation, and as such is vital to the country's development. Regulation of the mining sector has made rapid progress in recent years, with the exception of the environment. While the major companies in the mining sector already apply international environmental standards, the regulatory framework and institutions capable of ensuring that all companies are subject to the same regulations do not yet exist.

Moreover, there is no way of regulating either the artisanal sector or the rushes that occur every time a new gold or precious stone deposit is discovered, often in or near forests and protected areas.

In addition, according to article 35 of the new Protected Areas Code (COAP) [2008], for Natural Resource Reserves (category V) and Harmonious Landscape (category VI[147]) protected areas, mining and oil activities are permitted except in the hard core, subject to compatibility with the objectives of the protected area concerned.

In fact, protected areas are in constant danger because, in general, the operators' objectives will never be compatible with those of the protected area in question.

The *Mahafale* plateau is a veritable mining reservoir, home to resources such as bentonite, gold and iron.

During the transitional period up to the start of the Second Republic in 1975, the Malagasy government, with foreign aid (an American company), had already carried out an oil exploitation trial on the *Mahafale* plateau, particularly inside the current park. It was able to extract six barrels at the time.

As President *RATSIRAKA*'s regime opted for a policy of nationalisation and closure to Western countries, exploitation was halted. Currently, apart from the Malagasy government, the GEF of the World Bank and the German development bank KfW are co-financing the *Tsimanampesotse* park, while the French Global Environment Facility is financing the implementation of management transfers around the *Tsimanampesotse* park.

Donors are rushing to finance the park and its belt (TGRN). But what is motivating them to behave in this way? Is it because of the oil reserves in the limestone plateau and the various mining resources in the surrounding area, or some other motivation?

Whatever the reasons why donors are jostling for funding in the *Mahafale* plateau, this map below shows just how rich the protected areas are.

[147] The transfer of natural resource management falls into category VI

Map 10: Mines in WWF-managed sites, map drawn up in 2008

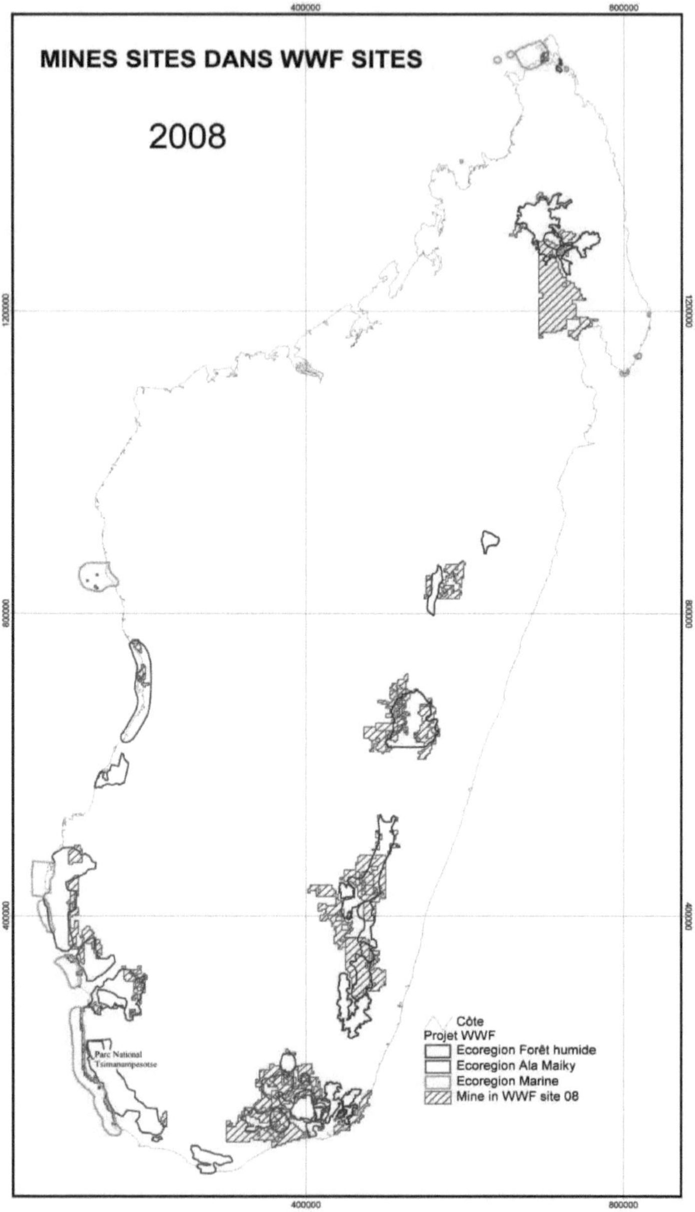

Source: WWF Toliara, 2013

There are various mines in the areas where WWF operates. They are not there by chance. In the case of the *Mahafale* plateau, which is home to the *Tsimanampesotse* park and the management transfers, the map shows us that in the south-south-eastern part of the new delimitation and in the following TGRNs of *Voroja* and *Ampitanake*, there is a trace of mining resources.

What's more, the Chinese are starting to approach the north-east of the park alongside the CoBa management transfers. Not everyone, not even the local authorities, knows what they are doing. They are used to clandestinely exploiting Madagascar's mineral resources. What's more, they neglect the environmental impact when they exploit mines. They have already exploited bentonite within the CoBa *Magnasoa Tany* management transfer in *Behomby*. This bentonite is used in oil drilling at *Sakaraha*. Neither the Forestry Administration Department (DREF) in *Toliara*, nor the *Beahitse* Rural Community, nor the *Behomby* CoBa managing the management transfer are aware of the Chinese exploitation and permit. It was the region's rural development department (DDR) that issued the permit. The DDR did not inform the Commune of *Beahitse* about the arrival of the Chinese, nor did it ask the local forestry administration whether or not the area in question contains protected areas. Like the Chinese, the Region's officials do not consider biodiversity conservation. While the State or the Ministry of the Environment and Ecology encourages grassroots communities to protect biodiversity, the Region ignores it. This proves that there is no communication between the authorities.

If the authorities who are going to enforce the law agree to be accomplices or plunder biodiversity outright, will the clearers and traffickers respect them or will they be afraid of them?

9.2 The future of forest resource management transfer in Madagascar

The majority of management transfers in Madagascar are established on the basis of management-exploitation and local development of biodiversity. Moreover, the initial text and spirit of the promoters of Law 96-025 (the Gelose Law) was to encourage the sustainable use of natural resources by communities secured by the transfer of management and recognition of exclusive land rights ("relative land security").

However, once the Gélose law was promulgated by the Malagasy government, numerous obstacles and even opposition delayed its application in the field. The body of foresters, identifying itself as the main manager of renewable natural resources, preferred the adoption of an application text to law 97-017 specific to the forestry sector. Decree 2001-122, known as the Gestion Contractualisé des Forêts decree, was promulgated as a complement to the Gélose law, but was seen as a de facto substitute [despite the strictly legal issues raised; Karpe, 2007].

In addition, the process of developing management transfers has been obscured, delayed and sometimes misguided, but it nevertheless seems to be developing inexorably to date. The creation of new protected areas as a result of the Durban Congress raises once again the question of the place given to local populations (inside or outside these resources?) and that of a still improbable synthesis between community management, sustainable management, the fight against poverty and the conservation of biodiversity.

Rural populations are therefore wary, because they feel betrayed so many times by the State and/or development and conservation projects. They believe that they are still living under a colonial policy of repression and exclusion from local resource management, even if the effective implementation of management transfer contracts is not abusive and, more importantly, the conversion of local daily practices towards sustainable management cannot be achieved without real and long-term involvement with them.

They feel excluded from their resources because the explanation of the different types of conservation that will be applied in Madagascar is not well received by the communities. They have confused "integral or conservative conservation" with "enhancement conservation".

Preservation by exclusion does not provide a real, lasting or serious guarantee of effective conservation of biodiversity. But denouncing irrational and destructive exploitation is not effective either if the needs of local populations are not met.

For example, according to Ravelona, the conservation management transfers put in place by Conservation International in the *Ambihilero* forest in *Didy* have never eliminated illegal logging in the areas placed under integral conservation [Ravelona, 2009]. This clearly shows that the creation of protected areas alone does not guarantee the sustainability of the biodiversity they contain.

Furthermore, government failures [Karsenty, Fournier, 2008] are often terrible for the biodiversity of protected areas. These failings can be caused either by a lack of resources on the part of the public authorities, particularly the government departments responsible for forests, or by periods of political instability. They do not make it possible to control flows or to combat illegal logging.

The conservation of biodiversity in protected areas is conditioned by two elements: the reality of the State and its ability to enforce the rules it lays down, and the reaction of the resident populations who have long occupied these areas and lived off these resources and biodiversity.

Periods of political instability, such as in 2009, are ruthless and exacerbated revelations of the failures of preservation in the face of more restrained but persistent and secretive practices in national parks and management transfers.

For more than eighteen years, the proliferation of the rosewood industry in Masoala Park and the exploitation of Prunus africana bark in the *Ankeniheny - Zahamena* corridor area (including the already existing *Zahamena* protected area and the *Ankeniheny* classified forest) have been held up as an example to looters and logging and mining companies.

Although Chapter V of Law 99-022 of 19 August 1999, entitled Mining Prospecting, stipulates that mining prospecting is unrestricted throughout the national territory outside protected areas, natural flora and fauna reserves and their protection zones governed by specific legislation, the proliferation of the barbaric exploitation of mining resources outside and especially inside protected areas is palpable. The rush towards clandestine exploitation of sapphires inside the *Zahamena* and *Isalo* protected areas bears witness to this. What's more, plans for a large-scale gold mining operation in the tapia forests of the *Soamahamanina* region are upsetting environmental protection fanatics.

If Madagascar wants to move towards sustainable management of its biodiversity, the political power that governs it must make a clear political choice that is accepted by the majority of the rural population.

9.3 Sustainable management and development or preservation of forest resources?

"Resaka tsy hay ny an'ny gidro sy ny ankoay ho an'ny fisa-tsinay ; Resak'adala izay tanety sy ny ala ho an'ny tsy misy drala" literally means that talking about lemurs and birds makes no sense to the hungry, and talking about hills and forests makes no sense to those who have no money. This reflection by the singer-performer Rossy illustrates the logic and perception of people who live in a heavy poverty from which they are unable to extricate themselves when it comes to preserving natural resources. It's hardly surprising that the prevailing attitude is one of 'give and take', as M.-L. Matthieu has shown in Mali [Matthieu, 2009].

Here, sustainable management refers to sustainable development[148] . Sustainable development is a concept that emerged from the movement for the sustainable management of natural resources, cited in the Brundtland Report [1987]. The report states that "*sustainable development meets the needs of the present without compromising the ability of future generations to meet their own needs*".

Sustainable development postulates that long-term development is only viable by reconciling the following three inseparable aspects: respect for the environment, social equity and economic profitability. The point of intersection and the main actor in these three aspects is mankind.

In concrete terms, sustainable development highlights the need to maintain or improve the quality of the natural environment, ensure the sustainability of resources, reduce differences in living standards between populations, promote the self-sufficiency of communities and enable the transfer of knowledge or wealth (including natural wealth) from one generation to the next. This concept is therefore concerned with the management of natural resources.

For the designers, Agar is a patrimonial management method par excellence, designed to give the people of a village back control over the ecological conditions in which they live. To implement agar, management transfer and community management are necessary. The latter are only instruments for achieving sustainable development, i.e. development, poverty reduction and sustainable local management of natural resources and biodiversity.

On the other hand, conservation drives people out of their environment. It will defend forests against people. In addition to the proliferation of illegal logging and degradation inside and outside protected areas, conservation is leading to an upsurge in rural banditry.

If we refer to the *Mahafale* plateau, it is preferable to briefly address this phenomenon of rural banditry.

9.3.1 The *malaso* phenomenon in the *Mahafale* plateau

It is no longer necessary to reiterate that cattle rearing plays a vital role among the *Mahafale*. Insufficient rainfall, poorly distributed throughout the year, has led to a lack of resources (fodder, watering points, land for cultivation, etc.) on the *Mahafale* limestone plateau. However, before talking about the *Malaso* phenomenon, it would be useful to talk briefly about the practice of transhumance and the traditions surrounding cattle rearing and transhumance.

[148] It is a gradual change that meets the essential needs of present and future generations by imposing limits on the use of precious resources and greater equity in the distribution of resources.

In the past, the practice of transhumance solved the problem of overgrazing, but with the scarcity of resources, aggravated by climate change and the increase in the population and ruminant livestock, conflicts have arisen between pastoralists and the local population in the destination areas.

It is to the centre of the *Mahafale* limestone plateau that the herds (almost all the cattle and a large proportion of the goats and sheep) move in order to take advantage of the pastures and water retained in the depressions during the rainy season. Herders practice transhumance at the start of the rainy season (November-December), which coincides exactly with the start of field work. The herds generally stay on their new pastures for 5 months, sometimes longer, returning to the villages from the end of April onwards over several months, with some herds not returning until June or July. It is important to note, however, that with the scarcity of grazing land, the transhumance movement is beginning to change in terms of duration and direction. The map below shows the flow of stolen zebus.

The comings and goings of the cattle during transhumance are very disparate, sometimes random. Some cattle are left in the forest by their owners for a number of days, but generally speaking, the main routes have been collected and referenced. In general, the route only changes when bandits are present.

9.3.1.1 Social interactions, conflicts and constraints

The *Mahafale* plateau is divided between major and minor clans, which have distinct influences on the territory. Several major clans and minor clans (in the *Ankazomanga* area) are present on the plateau. The clan areas are well defined[149] and separated by corridors of "no man's land" (no clearly defined "owners").

The delimitation of grazing land in the rainy season is done by agreement between the clans of the coastal plain and those of the inland *Mahafale*. Transhumance routes are therefore linked to the presence of "allies" in the host areas. In principle, these alliance networks - made up of matrimonial ties, *ziva* (joking kinship) and *fati-drà* (blood brotherhood) - were concluded between clans of the same *Mahafale* royalty (*onilahy, linta, menarandra*). For each clan on the coastal plain, the transhumance route and locations have long been fixed and are more or less the same every year. The right to graze on the territory of the inland clans seems to have been granted without the users of the coastal plain being obliged to pay them any fees for the grass consumed by their herds.

9.3.1.1.1 Polygamy and the movement of oxen

Polygamy, a traditional practice among the *Mahafale*, often serves political ends. Originally, the matrimonial alliance was motivated by the need to create a climate of trust and good understanding in order to gain access to pastoral areas. In social terms, the wife guarantees the continuity of the group. When a man takes a new wife, he must compensate his first wife with an ox. In return, his parents-in-law give him a *tandra* (dowry) - a cow. This *tandra* bears the ear-mark or *vilo* of the wife.

[149] Although we did not manage to define them precisely during our surveys

9.3.1.2 Current conflicts

Although it is difficult to "measure" the evolution of trends, as we have no reference other than the memories of the "elders", the majority of players testify to a gradual disappearance of the application of and respect for traditional customs[150] . It would appear that these rules have been in disuse for some twenty years. "*Transhumant herders are not the same as they used to be, they no longer respect the rules we have, such as preventing zebus from entering crop fields"* [Pastor of *Etrobeke*].

The main "objective" facts that demonstrate the evolution of practices are as follows:

(i) the abandonment of certain traditional rituals such as the *titike*, a social pact between transhumant herders and local people during which the rules for using the land were defined;

(ii) the direct installation of herdsmen in pastures without prior negotiation with the locals;

(iii) non-compliance with rules by transhumants in certain host localities (non-compliance with taboos around watering points, non-compliance with zones defined for cultivation and grazing, etc.);

(iv) disrespect for and/or distrust of traditional (*Kabary*) and modern justice systems;

(v)the transition from "contemplative" zebu management to more "capital-intensive" management.

9.3.1.3 Zebu theft

Like the turning of the dead in the highlands, zebu theft is considered a tradition in southern Madagascar. It will never die out. Nevertheless, there are periods when it is on the increase and periods when it has subsided. Pavageau has noted that *"zebu theft, like the turning over of the dead, has always been presented as a specific feature of Madagascan culture, a curious and exotic practice that continues to this day"* [Jean Pavageau, 2008].

[150] 90% of the *Mpitan-kazomanga* surveyed said they had abandoned traditional management rules

Map 11: Pastures, water points, transhumance routes and *malaso* routes

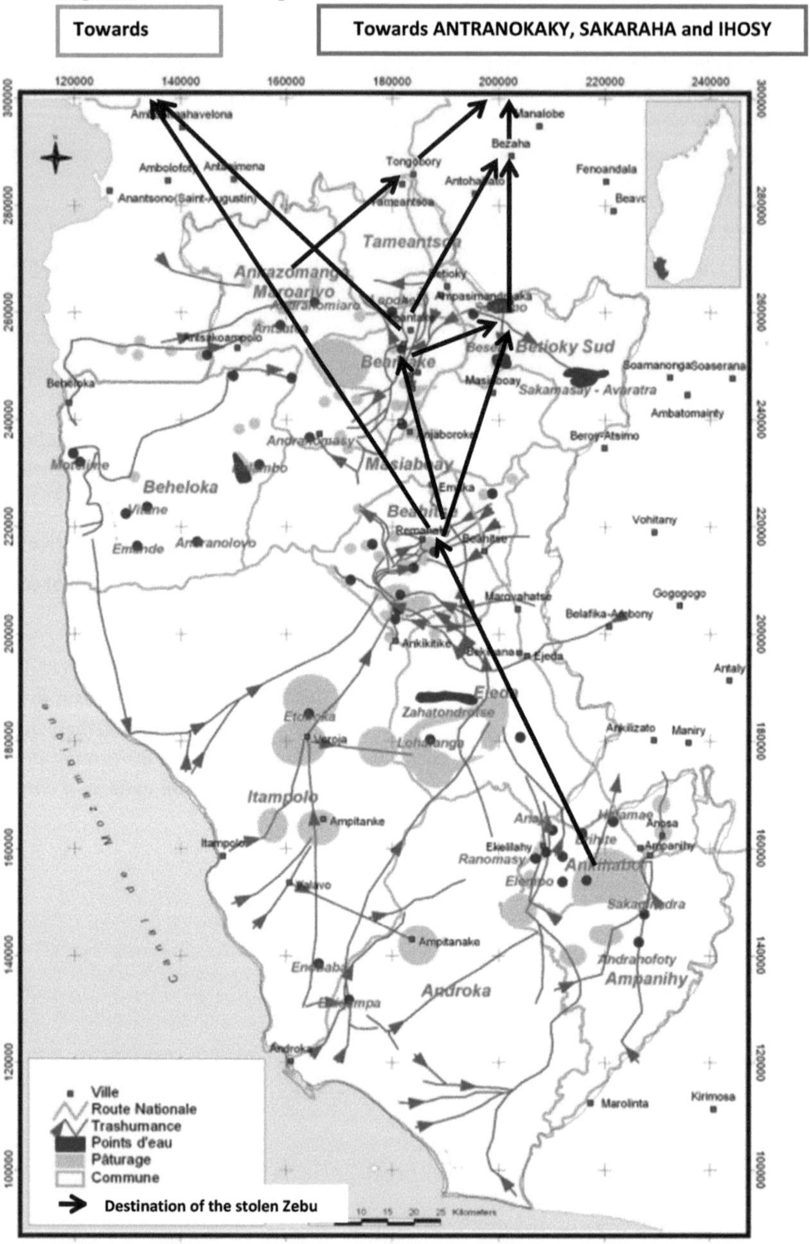

Source: WWF Toliara, 2014, improved

9.3.1.3.1 History of beef theft in Madagascar

The *malaso* phenomenon has existed for a long time. At the time of the *vazimba* society, there was already talk of *sonjoa*[151] [Rasamoelina H, 2000]. During the feudal period, particularly during the reign of Ranavalona II, those who were brigands and thieves were called *fahavalo* or enemies. The name *fahavalo* was also used during the colonial period to designate resistance fighters. The export of zebus stopped with the abolition of slavery by the colonisers. It was they who exploited this industry afterwards.

"On the other hand, Mr. Augagneur believes that Madagascar's commercial future lies with South Africa.

This is why we have completely rebuilt the western route, equipped the port of Tuléar and created a regular coastal shipping service on the west coast. Madagascar can supply South Africa with livestock, timber, rice and coffee, and the return freight will consist of coal, which the island badly needs. [Zimmermann Maurice, 1909]

If we refer to these texts by Zimmermann, the export of livestock was already practised during colonisation. During that time, it was the colonial administration that managed exports. In other words, it was the colonial administration that negotiated with foreign countries on the goods to be exported. The practice of exporting zebus was therefore inherited from the colonialists.

At the start of independence, the colonists continued to export zebu cattle to South Africa and the Indian Ocean islands [the Comoros and Reunion archipelagos].

The settlers installed collectors at the collection points. In the case of the *Mahafale* plateau, *Betioky* was one of these. To ensure the expected weekly quota or to facilitate the collection of zebus, the settlers installed collectors (for the *Mahafale* plateau, there was one person from the south-eastern part of Madagascar, *Farafangana* to be precise) in strategic locations. The collector sent the cattle to *Toliara* when he was able to gather the number of cattle required by the settlers. Once the zebus had been assembled in *Toliara*, they moved on to export at the port.

It should be pointed out that for the *Bara,* this used to be seen as a rite of passage into adulthood. To prove their virility, or to prove that they could support a household, young boys had to undergo this "ox-stealing ordeal".

During the Second Republic, the State [or the Head of State?] took over from the colonists to ensure the export of zebu to the Indian Ocean islands. It was at this time that the "*dabok'andro*" (zebu traders) came into being. The *dabok'andro* are responsible for buying zebu for resale to collectors. They are also paid for transporting the zebu to its destination. To ensure the due date and to guarantee the collectors' order, the *dabok'andro* have brought in ox thieves.

Since the 1990s, stealing oxen has become a "profession" for the *Malaso.* As a result, their cattle-stealing practices and strategies have changed, as they too have a deadline to meet.

In recent years, especially following the political instability of 2009, there has been an upsurge in theft and violence in the *Mahafale* Plateau area.

The destination for zebu exports is no longer the Indian Ocean islands, but China. What's more, collectors have raised prices, because a castrated zebu aged 3 years (*telo ay*) or more costs 3,000,000 *ariary* and those under 3 years old 1,000,000 *ariary* per zebu head. The young people

[151] They are thieves of cattle and human beings whom they reduce to slavery

of the *Mahafale* plateau are tempted by this price. They don't give a damn about the *titike* and a collective agreement or "*dina*" sometimes called "*dinan'ny mpihary*" established by the communities.

9.3.1.3.2 The organisation and strategy of *malaso*

Attracted by the exorbitant price of stolen livestock so that they can quickly change their social rank, young boys take up *malaso* at the age of 14. They have left farming behind. They no longer clear forests to grow crops. They have grouped together by village. Neighbouring villages cannot attack each other directly. However, they will guide other *malaso* who come from afar to attack neighbouring villages. In fact, each group must ally with the other groups on the other bank of the *Onilahy* (*Bezaha*) to ensure the transport of oxen.

After the *Remenabila* phenomenon, during each raid, the number of *malaso* varies between 100 and 300 people (men and women). This is known as *Barinjaka*. The women who accompany the *malaso* during the raids are called *Baraondry*. There are between 3 and 6 of them, and they sing and loot kitchen utensils, mattresses and blankets from every household in the village that is attacked.

Nowadays, especially after the practice of dina, the *malaso* use a new attack strategy. They will take the owner or his close family hostage so that the targeted zebu owner will negotiate the ransom with them (zebus or money and a rifle).

The *malaso* camp around the transhumant reception areas. However, they do not attack the transhumants during the period when the latter remain in the reception area because the oxen cannot cross the *Onilahy* river. The transhumants are in the reception areas during the rainy season and they cannot cross the *Onilahy* because the river current is very strong. So they wait until May or June to attack. They steal the oxen, kill the villagers and burn down their houses, and this has not stopped despite the increase in military operations.

Toliara or *Antranokaky*, *Sakaraha* or *Ihosy* are the embarkation and transaction points. The *malaso* are obliged to collaborate with the *ombiasa*, elected representatives [such as the mayor and sometimes the deputy] and public administrations such as the arrondissement delegates, and the gendarmerie to settle the paperwork in good and due form [the individual bovine file known by the acronym FIB and the bovine notebooks (*bokin'omby*)]. As for the *ombiasa,* they play a very important role for a group of *malaso* because they act as a "coach" responsible for the group; they ensure the "grigri by bullet" during an attack and guess the best days to organise the attack. Together with the *malaso* group leader, they organise the routes, the sentries, the intelligence agents and so on.

For the destination of *Toliara,* they have to settle the paperwork for the stolen cattle at *Manorofify* (the commune downstream from *Onilahy*). For those who choose *Ihosy* or *Sakaraha*, the paperwork for these cattle is settled at *Bezaha* (the commune upstream of the *Onilahy*).

Farmers will lose hope if the tracks of their cattle are lost on the banks of the river *Onilahy*. Once the cattle arrive in *Manorofify* or *Bezaha*, they write down and describe the colours and new signs marked on the stolen zebus in a new cattle book pre-filled in someone else's name.

It has to be said that ox thieves do not attack the pastoralists on the coast because they cannot find ways to take their cattle away. However, at least three cases of cattle theft a week are reported in the "*fatrambe*" (or inland) *Fokontany*. The minimum number of stolen zebus is two (cart puller).

In general, stolen zebus will not be recovered if one of the *malaso* members does not betray his members[152] or if the stolen zebus belong to a close family member (uncle) of a *malaso* member. In this case, the family member who has lost his cattle must pay a sum of money to the *malaso* who are holding the cattle.

9.3.1.3.3 Agropastoralists' strategy in the face of *malaso* attacks

On the *Mahafale* plateau, agropastoralists still maintain a customary system of village alliances and territorial surveillance designed to combat theft by detecting animal movements. Each village is responsible for guarding its own territory, which is based on monitoring the tracks left on the ground by the zebus at strategic crossing points, the *kizo*, which are watched over and swept up daily.

It should be noted that agropastoralists on the *Mahafale* plateau generally choose to breed red-coloured zebus. The reasons for this are, firstly, that zebus of this colour are easy to sell, as the *Mahafale* are beginning to show an interest in them; secondly, to distinguish themselves from other agropastoralist territorial groups such as the *Antandroy* and *Bara*, and also to try to control the traceability of the zebus.

According to local perceptions, the theft of cattle on the *Mahafale* plateau is a source of social conflict. Suppose zebus have been stolen and lost from the tracks in any village in a Commune, if the tracks have not left the village, all the people in the *Fokontany* are responsible for replacing the lost zebus. In the *dina*, one lost zebu will be replaced by four other zebus. But above all, the village where the tracks were lost must take over the task of following up the tracks of stolen zebus (*magnara-dia*). If the people in the village where the tracks were lost refuse, they must replace the lost zebu following the *kabary ambany kily*.

In recent years, farmers have developed strategies to deal with theft:

a) keeping cattle in pens inside villages at night.

b) transhumance to the coast in the dry season, as theft of zebus is rare in this area,

c) a reduction in savings in the form of zebus, offset by increased investment in land (purchase of land).

Also, some households in the Communes are equipped with at least one shotgun to protect themselves against thieves and ill-intentioned people. Some households even own pistols to ensure the safety of their families. Apart from that, to defend themselves, the population had applied village self-defence.

From 2012, villages that are able to contribute the sum of 400,000 *Ariary* to pay two soldiers will apply to the Chief of Staff of the military army in *Toliara*.

At present, the *dina be*, the last resort found by the *mpitan-kazomange* in the *Ampanihy* and *Betioky* districts, has been approved by the *Toliara* court after being requested by the Head of Region. The latter is convinced of the effectiveness of the *dina be* and is going to implement it

[152] Betrayal between members is caused by internal conflict. However, most malaso practise blood brotherhood or fati-drà to expand their network because a fati-drà of one *malaso* will automatically be fati-drà of the other fati-drà even if they do not drink each other's blood.

throughout the region. It should be noted that after the implementation of *dina be* in the *Mahafale* plateau by the *mpitan-kazomanga* of the two initiating districts, *malaso* attacks have decreased. For the moment, we do not know the cause of the reduction in attacks, whether it is really the effectiveness of this *dina* or the non-existence of stealing oxen in the *Mahafale* plateau.

9.3.1.3.4 Why is beef theft on the increase?

In general, when grassroots communities disagree with the central state, the *malaso* phenomenon appears. Rasamoelina H, in his study entitled l'Etat, les communautés villageoises et les phénomènes pathologiques en milieu rural malgache : le vol de bœufs et les feux sauvages, noted that the *malaso* phenomenon first reappeared after the establishment of the SMOTIG, or service de la main d'œuvre d'intérêt général, in 1927, more precisely in 1930, and then after the price was fixed and the Office du riz established in 1943. It did not end until the mid-1950s. As we know, however, rural insecurity reappeared in the early 1970s[153] , only to worsen and become more widespread from 1980 onwards, with a certain lull in the mid-1990s [Rasamoelina H, 2008]. After 2009, the situation worsened still further because the *Malaso* are well equipped[154] and behave like militias. When they attack a village, their numbers must exceed 40 men.

Furthermore, the habits and customs governing cattle rearing seem to be outdated. Most transhumant herders no longer practise these traditions, perhaps because of the entry of development projects into the *Mahafale* plateau over a decade ago. These development projects generally work to "de-culturise" the *Mahafale* so that they can implement the development model desired by the donors.

The phenomenon of migration that is currently afflicting the plateau is the result of this *Malaso* phenomenon, as those who no longer have oxen leave their native land to live in the north of Madagascar (*Ambilobe*) or *Morondava* or in the sapphire quarries. They are a long way from *Tsimanampesotse* Park and the forest resources that donors have described as being under anthropic pressure.

In addition, the *malaso* phenomenon today has an economic objective. It has always been very difficult to assess the cattle population on the *Mahafale* plateau. First of all, and although this may seem contradictory, large-scale farmers prefer to conceal the size of their herd, even if their prestige remains proportional to the number of zebus they own. Secondly, the introduction of taxes on cattle by the colonial authorities contributed greatly to the underestimation of the herd.

By calculation or not, after the failure to set up the BTM bank (today's BOA) in *Ampanihy*, which closed its doors in the 1980s, and after a series of failures to set up microfinance in the *Mahafale* region, they were able to set up a microfinance institution called FIVOY in 2014 because of the decline in cattle numbers in the region.

Today, an accurate census of herds seems to be one of the most effective ways of combating theft, but the proliferation of stolen zebus still leads to concealment.

9.3.2 Sustainable management or preservation of forest resources?

---

[153] In the deep south, this phenomenon began to worsen after the insurrection led by Monja Jaona in 1971.
[154] The *Mahafale* plateau has now become a destination for trafficking in automatic weapons.

For many years, the conservation of biodiversity has been at the centre of vast debates, both political and scientific. The opposition between these two visions of conservation, sustainable development and biodiversity management, fuels the discussions.

This influences the actions of non-governmental organisations working in Madagascar. On the one hand, they advocate strong protection of the planet's last remaining areas of significant biodiversity, generally implying a halt to all human activity that excessively consumes natural resources, and on the other, greater recognition of the role of local populations and the importance of taking them into account in decision-making.

At international level, the contradictory debate between conservation and poverty reduction remains persistent, and Madagascar has not been spared as an example. In Madagascar, a long period of almost a century of repressive and exclusive forestry policy came to an end in the 1990s, but the change has not yet been lasting, as most of the decrees applying the Gelose law or the GCF decree did not exist. There is no guarantee today that Madagascar will continue its slow progress towards sustainable development.

The contradiction between conservation and community management (i.e. conservation, sustainable management and development) really disorients the debate on sustainable management. Safeguarding Madagascar's biodiversity cannot be resolved and limited to a network of protected areas, however vast and impenetrable. Sustainable management of all natural areas is the only way to guarantee the effective conservation of exceptional biodiversity. This sustainable management throughout Madagascar cannot avoid involving rural communities and communes, through contracts based on the models initiated by the "Gélose" law.

Barbaric conservation[155] " is leading to an upsurge in illegal logging of forest resources. The State must dare to choose between the policy of community management and that of "barbaric conservation" of renewable natural resources, which excludes man, an element of biodiversity. It must not allow itself to be influenced by the various theories and experiments of rich countries.

9.4 **Suggestions and proposals for improving community management**

The former Gélose areas are being transformed into GCFs because the land is too large, the SFR has failed and there is no environmental mediator. Although transfers of renewable natural resource management decreased after the extension of the *Tsimanampesotse* park, they are still undergoing this transformation. The implementation of a large-scale project such as the extension of a national park or the creation of a new protected area must include appropriate accompanying measures.

9.4.1 Drawing up a *dina be-type* environmental agreement

The illegal and clandestine exploitation of natural resources, especially endemic ones, continues in Madagascar despite the abundance of protected areas. Looters are not shy about entering protected areas. All the neighbouring communities have seen the Malagasy trees being shipped off to China, but they are keeping quiet. They are waiting for the State to react, perhaps so that they don't "chew before their molars" *(mitsako alohan'ny vazana)*. They are ready to protect the

[155] In this case, barbaric conservation means conservation without concern for the communities that depend on these resources.

surrounding natural resources if the State shows an interest in preserving them. The Malagasy government must therefore work with grassroots communities and civil society organisations working in the environmental field to draw up a *dina* that will put an end to the problems of illegal logging. This *dina* will apply to everyone without distinction in Madagascar.

9.4.2 Generalisation of the land tenure security process

Concerted securitisation should not stop at this method; it is recommended that concerted securitisation be included in the study currently being carried out by the recently established committee for the revision of old texts. This insertion concerns the proposal for laws on access to indigenous resources, even though it concerns the GCF. In this case, CoBa land will be considered as having specific land tenure status.

The methodology of concerted securitisation has a very conceivable perspective. It can be used and developed as part of a project. It is possible to create a logical matrix model (based on the logical framework method). The heuristic diagram enables the strategic objectives to be set and the main evaluation criteria to be drawn up. The logical framework can also be used to identify monitoring indicators, estimate costs and set work schedules. This methodology can be implemented by various bodies, but it is preferable that it be supervised by the Regional Department of Ecology, Environment and Forestry. In any case, it is the public meeting that decides.

The various mines on the *Mahafale* plateau have gradually come to light, although the mining code gives priority to the exploitation of these resources regardless of their location. The areas transferred to the CoBa need to be secured through the implementation of the SFR, as the managing CoBa have already spent a lot of money to protect them.

What does the future hold for the TGRNR in the face of the rapid expansion of industrial mining operations and the Olympian publicity surrounding the commercialisation of land by the State through the guichet foncier? If they were set up to protect the parks, don't they deserve the same security equipment as the parks?

9.4.3 Maintaining a close relationship with the forestry administration

The re-delimitations organised by the WWF have led to zoning operations. Given that the former Agar zones have been mutated into GCF zones, the names of the zones have changed. This change is due to the fact that the GCF only concerns forests. All other types of land use must be excluded. Following the failure to set up SFRs under the Agar system, the TGRNRs under this system in the nation will mutate to the GCF system. In this case, zoning should only concern forests.

The definition of zones in the TGRNR explains the objectives of exploiting resources in different areas. This principle is based on the consideration that all transferred areas are forests. "Forests, within the meaning of law no. 97-017 of 8 August 1997, are defined as all areas meeting the following qualifications: areas covered with trees or woody vegetation, other than those planted for the exclusive purpose of fruit production, fodder production or ornamentation; and areas occupied by trees and bushes located on the banks of watercourses and lakes and on eroded land"[156] . Article 2 of the same law stipulates that non-wooded land, land cleared without

[156] Art1, law no. 97-017 of 8 August 1997, revising forestry legislation, decree no. 2005-849 of 13 December 2005, J.O. no. 2449 of 25.08.94, p. 1717.

authorisation or clearings are forests. Formerly cultivated areas are formerly wooded land that has generally been cleared for less than five years. Hence the name agro-sylvo-pastoral zone. The other areas are for forestry purposes and are intended for the regeneration, restoration or enhancement of species. These are the conservation and use right zones. The TGRNR is therefore based on a very broad legal definition of forests, which nonetheless contradicts the practices of local people who appropriate farmland by clearing it.

As a result, maintaining relations with the forestry administration seems to be a guarantee of sustainable exploitation of the resources, but explicit land rights should be recognised for the local populations on the land included in their terroir.

In addition, the implementing regulations for the Agar Act and the GCF decree need to be completed to ensure their successful implementation.

9.4.4 Adapting the management transfer to existing customs and practices

Madagascar is a country where the cult of the ancestors still reigns. It is necessary to reiterate that in the Gelosa concept, community conservation areas extend further than resources and include part of the village land governed by communities. Everywhere in Madagascar, natural resources are intact because they are protected. They are protected either by forest spirits or by the community itself through traditional chiefs. They belong to those who protect them.

The traditional chiefs, generally the elders, are the point of reference when it comes to respecting traditions, habits and customs. They hold the *hazomanga*, the symbol of "wisdom", and help the *fokonolona* make decisions on matters that concern village communities alone. For the management of these resources to be sustainable, the initiators of the management transfer project must strongly involve the traditional chiefs and *ombiasa* so that they can join the ranks of the management committee, the managers of these resources. This means taking into account the traditional forms of management that the communities have already applied to manage the forests in their territories.

Support measures will be created for grassroots communities, such as functional literacy and income-generating activities. The development organisation must commit to supporting income-generating activities to ensure the viability of the TGRN; these activities must be identified when the development plan is drawn up.

Otherwise, the support organisation and the forestry administration must sign a collaboration agreement, either overall or by TGRN, in which they reciprocally undertake to support the transfer beyond the signing of the contract, and to ensure the succession of financial support by linking the CoBa with development programmes.

Finally, taking into account the aforementioned uncertainties, the outlook for CoBa adaptation should be based on three elements:

- the stability of the areas/terroirs of the transferred forests, to enable the establishment of a long-term adaptation strategy based on trust in the administration ;
- the inclusion of land issues in written *dina*, in order to promote the securing of land rights recognised by the CoBa and their members, and
- the implementation of a social safeguard plan for the management of the park and its surroundings to support the CoBa in transforming their farming practices, made

necessary by the reduction in access to resources.

Conclusion part 3

To conclude, this section argues that the Gelose approach is modelled on colonial forest management, whereas grassroots communities have long had a traditional form of management that favours free access to natural resources. As a result, this new approach has a number of implications and limitations, given that the context and implementation period are different. Those involved in the conservation of Madagascar's biodiversity each have their own objectives. This leads to differences in the level of involvement of each. The degree of involvement of the State and the forestry administration in environmental protection, for example, is low. We can see this situation in the failure to take decisions on certain texts applying the Gelose law.

Following the proliferation of slash-and-burn agriculture, the remnants of the forest on the *Mahafale* limestone plateau are taboo and/or sacred forests. These are demarcated and form part of the management transfers. Previously, the forests were managed by the spirits known as *Tambahoake*, transferred to the *ombiasy* and *mpitankazomanga*. Today, these forests are devolved to local communities run by management committees (ordinary citizens who are neither *ombiasy* nor *mpitankazomanga*). It is not thanks to the Gélose law or the GCF decree that these forests have been preserved, but rather thanks to the habits and customs that the *Mahafale* still respect to this day.

The aim of NGOs and CoBa in conserving biodiversity is to make money from it: funding. However, they are convinced that the conservation of these forest resources will not last, because since the early 1980s, the *Mahafale* plateau has been going through an unprecedented ecological crisis, resulting in intense deforestation due to maize farming. What's more, the management transfers put in place around the park are a long way from the hard core to ensure security.

As a result, natural disasters are difficult to control (cyclical droughts, locust invasions), the crisis in the rural world, which is largely manifested by insecurity, the fall in agricultural production, the downward trend in cattle numbers, and population growth leading to increased demand for forest products (charcoal, firewood, schooners). All of these factors unfortunately contribute to the spiral of degradation of the natural forest resources of the *Mahafale* landscape.

Forest exploitation (deforestation and use of other forest products), one of the activities of last resort and a source of income, is part of a strategy that enables people to survive as best they can and to accumulate funds that will allow them to re-establish their image in society by acquiring a large herd of cattle.

The threats to biodiversity on the *Mahafale* Plateau are, in order: clearing for slash-and-burn agriculture for local consumption; clearing for maize slash-and-burn cash crop agriculture; clearing for charcoal production, collection of forest products for fuelwood and construction, bush fires, collection of non-timber forest products; population growth and mining.

GENERAL CONCLUSION

The participatory method is used in the implementation of the Gelose and GCF decrees to conserve and/or manage Madagascar's biodiversity. In addition, the study of citizen participation in the process of transferring the management of renewable natural resources enabled us to observe the cleavages in the *Mahafale* plateau, the power relationships between the active people and the rest of the population, and the latent conflict between the indigenous people and the migrants over the management of forest resources.

In short, at the end of this analysis, the results will reveal that this study has helped to highlight the shortcomings of the texts implementing the Gelose laws and the GCF decree and the causes of the non-participation of the citizens concerned in decision-making and in the management of the transferred ecosystem. However, the results of this study are not definitive, but should serve as a basis for future research, as social reality is in a perpetual state of flux. Moreover, it will not be possible to assess the impact of Agar for less than 25 years (one sociological generation), because Agar or the GCF decree has a long-term impact.

Our observations and interviews on the *Mahafale* plateau have shown that most of its inhabitants seem apathetic or incapable of fulfilling their duty to society. Even though they generally recognise the benefits of conserving forest resources, there has been no real participation from everyone in the gathering of their aspirations, in the implementation of the management plan, or in the other actions to be carried out within the CoBa.

Identifying the beneficiaries of the management transfer, on the one hand, and the needs of a given population, on the other, always poses a problem. Logically, the beneficiaries should be those who express their needs, i.e. the entire community. But in the context of the *Mahafale* plateau, those in need seem to be leaving the choice of projects or activities to the leaders (CoBa management committee, forestry administration, NGOs). However, they are going to suffer because the aid or funding for the protection of the forest ecosystem will not correspond to their needs. The mobilisation of future beneficiaries is still ongoing. The adoption or appropriation of the management transfer by local people will depend on the satisfaction they feel.

The presence of forests that are still preserved is the result of a combination of natural factors (difficult to access and therefore not very integrated into the monetary economy, poor soil unsuitable for farming, available and exploitable agricultural areas all around that still protect these forest blocks, distance from the forests, water problems), demographic factors (low human density or balance between population and resources still maintained), socio-cultural (presence of well-respected *"faly"* taboos, *"ombiasa"* diviners and healers, guardians of the sacred and on whom the cohesion of the group depends, *"mpisoro"* or *"mpitankazomanga"*, guardians of the *"hazomanga"* or tutelary poles) and organisational (presence of a homogeneous social structure that has been little disrupted by migrants and of a well-structured social organisation managed by notables, who have great respect for the forest and who are both wealthy breeders and large landowners).

Lack of water leads to health problems and malnutrition, and becomes the primary concern of communities, both for their survival and for their livestock. The aspirations and activities of the inhabitants depend on it. During the dry season, people have to walk several kilometres to get access to non-potable water, which is used for both human and animal consumption. The absence of water leads to uncitizen practices and the degradation of biodiversity. As a result, the issue of conserving renewable natural resources is being called into question, as one of the communities' basic needs has yet to be met.

Farmers on the *Mahafale* plateau seem to have long experienced the destruction of their maize fields by nomadic and migratory locusts. From 1995-96, the locust invasion became a scourge, destroying a large part of agricultural production, especially cereals, in a region where food insecurity is becoming chronic.

It should be remembered that our first hypothesis seeks to verify that the level of participation of the grassroots community depends on its level of education, and non-participants at the lower end of the scale do not want to compromise themselves in matters that they consider obscure and dangerous. Members of the *Mahafale* community who are illiterate do not want to compromise themselves in the management of these natural resources compared with those who have an average level of education. The latter want a new status that will help them dominate the community, so they join environmental and development NGOs. They try to impose themselves on the members of the entire community by concealing information about the management transfer and the CoBa. As for the illiterate, they don't know much about what is meant by a transfer of management, they think that the land of their ancestors, including the forests, has already been sold to foreigners, and they don't dare join CoBa because for them, being a member of CoBa means being an accomplice in this sale of the land of their ancestors.

In addition, the grassroots communities managing the transfer of natural resource management on the *Mahafale* plateau have a dual objective: to be a model for other communities and also to benefit from funding to meet their basic needs. These communities are unable to manage their natural resources properly because of a lack of funding. All the players involved in this management transfer are simply confiscating traditional management by the *ombiasy* and *mpitan-kazomanga*. The success of this management transfer has long been based on respect for customs and traditions, but above all on respect for the people of *Tambahoake,* the guardian spirit of the forests of the *Mahafale* plateau.

Due to the reduction in the communities' rights of use, the geographical situation and climate change, the objectives of sustainable management of natural resources will not be achieved because the accompanying measures for the benefit of the communities of the *Mahafale* plateau have not been put in place and/or implemented. The proliferation of illegal trafficking in natural resources is increasing and land clearing is on the increase in the management transfers within the *Tsimanampesotse* Park.

At the end of this analysis, the results discussed in this book allow us to draw the following conclusions on the conditions that could lead citizens to participate more fully in the conservation of biodiversity in their region:

- Firstly, the integration of the underprivileged classes as well as migrants stimulates the participation of local residents in the management of forest resources; family ties between all residents and their enhancement accelerate the effective mobilisation of citizens;

- secondly, it should be pointed out that trust and dialogue between the forestry administration, the management committee and the communities form the necessary basis for carrying out the powers and duties of the CoBa. Thus, the socio-political relationship in the

Mahafale plateau would favour the participation of all citizens if the various economic or environmental associations created in the commune really facilitated collective life and ensured the regulatory control of some members of the community over the agents of power;

- thirdly, management transfers in the *Mahafale* plateau must be "mahafalised" if the plateau's natural resources are to be conserved sustainably. "Mahafalisation" of management transfer means adopting traditional management based on respect for customs and traditions. The people primarily responsible for transferring management are traditional chiefs such as the *"mpisoro"* or *"mpitankazomanga"* and the *"ombiasy"*. Customs and traditions concerning forest management must be taken into consideration in areas where the Gelose law and the GCF decree exist;

- Finally, alternative activities more suited to the climatic conditions of the *Mahafale* plateau need to be carried out. The introduction of direct-seeding cropping systems, which protect the environment and regenerate soil fertility, and the use of drought-resistant seeds will help the communities living alongside the park.

In fact, residents' participation depends on their belonging to social strata or groups born of real cleavages within the community and virtual cleavages between members and leaders within formal and informal social groupings and outside these same entities. So the absence of real cross-fertilisation within the community is a barrier.

Traditional social organisation and customs ensure the stability and success of management transfers in the *Mahafale* plateau, this was the second hypothesis that we verified. The *Mahafale* are very attached to customs and traditions. They are the most traditional of Madagascar's territorial groups. Customary rules form the basis of *Mahafale* social organisation. The sustainability of traditional management and the use of forest resources are guaranteed by these rules. The secret to the success of the transfer of management of the *Mahafale* plateau is respect for customs and traditions by the communities, as most of the forests transferred to the grassroots communities include sacred or taboo forests. Our ancestors already managed these forest resources. Their management was based on dissuasion through threats, participatory management and raising awareness, showing that the forest is ultimately the last resort of the poor.

So the policy of participatory forest management is nothing new. The way in which it has been implemented has changed from one regime to another. At present, the transfer of natural resource management has taken place thanks to the implementation of the Gelose law and the GCF decree. Donors have entrusted both national and international non-governmental organisations with their implementation.

Indeed, the conservation of renewable natural resources is becoming a real business, a real international competition between NGOs. It is not the natural resources that concern them, but the market. In the case of the *Mahafale* Plateau, the intervening NGOs do not have the same procedure despite the coordination efforts undertaken. The divergent approaches of the technical and financial partners mean that the support organisations are weak, even though they are working on the same issues and in the same sector of activity. This situation is deplorable insofar as local grassroots communities take advantage of it to obtain support that does not always meet their real needs. This makes it difficult to achieve effective ownership and sustainable results.

Finally, poverty is a source of apathy. It does not allow the poor to be aware of the duty that awaits them. The poor's strategy for getting out of their situation is to join the CoBa in order to survive. In this way, the poor take part in work that requires physical effort, with varying degrees of remuneration, while decisions are left to the rich. The needs and aspirations of the poor have always been neglected to the detriment of those of the rich. What's more, these poor people do not have sufficient technical skills. They are also ill-informed, and the most active even try to hide this fact, while a minority try to involve them in the conservation of forest resources.

Our last verified hypothesis is that the management transfers in the Mahafale plateau do not achieve the objectives of sustainable management because of the reduction in the communities' use rights, the geographical situation and climate change, which weigh heavily on their land. The *Mahafale* plateau is hit every year by food insecurity, which has encouraged abusive exploitation, and by the insecurity of the communities in their current management methods. This food insecurity is exacerbated by the vagaries of the climate and the locust invasions that plague the area every year. To overcome this, the inhabitants along the RN10 have turned to charcoal-making to survive. What's more, the *Mahafale* communities have no alternative source of income, as the support projects are not doing enough to promote alternative income-generating activities that respect the environment.

Despite the persistence of this chronic malnutrition, demographic growth is a recent phenomenon in the *Mahafale* plateau, and this has an immediate impact on the dynamics of natural resource management. It has led to the saturation of the low-lying valleys and *baiboho* that can easily be developed, and has also triggered major migratory flows to forest areas with little human settlement. It is also forcing the conquest of new land to be developed at the expense of the forest. The needs of these inhabitants have increased following this demographic growth, and the anthropic pressure on natural resources has therefore increased.

From what we have been able to study, the three conditions that encourage the population to take part in natural resource conservation through management transfer are a relationship of trust between the leaders and the led, the integration of the community and poverty or its education.

So, in order to conserve Madagascar's renewable natural resources, and given the example of the *Mahafale* landscape, an in-depth study of the culture is necessary to obtain greater participation. This would enable us to get to know the people of the area better and promote communication and, above all, awareness-raising among the population.

BIBLIOGRAPHY

GENERAL WORKS

1. BALANDIER (G), "*Sens et puissance : les dynamiques sociales*", Paris, Edition PUF, 1971.
2. BALANDIER (G), "*Anthropologie politique*", Edition PUF, Paris, 1967.
3. BALANDIER (G), "*Le pouvoir sur scène*", Edition PUF, Paris, Fayard, 1992.
4. BOURDIEU (P), "*Sur le pouvoir symbolique*", in Annales, Economies, Sociétés, Civilisations. 32è année, N.3, 1977, pp. 405-411.
5. BEAU (M.), "*L'art de la thèse*", La découverte, Paris, 1985.
6. CALLET (R.T) " *Histoire des Rois (Tantaran'ny Andriana)* " translation by CHAMPUS (G-S) and RATSIMBA (E), Librairie de Madagascar Tana 3 Tomes Pasteur.
7. CROZIER(M) and FRIEDBERG (E), "*L'acteur et le système (les contraintes de l'action)*" édition du seuil, Paris France 478p, 1993.
8. CROZIER (M), TROSA(S) (dir.), "*la Décentralisation, Réforme de l'Etat*", Editions pouvoirs locaux, 1992.
9. DURKHEIM (E.), *"Règles de la méthode sociologique"*, PUF Publishing, Paris, 1977.
10. EDWARD (B. T.) *" La sociologie : dictionnaire du savoir moderne"*, tome 1, Ed Gérard &C°, p71
11. FAUCHER (D.), "*Géographie agraire*", Paris, Génin, 4th Ed, 1979
12. FAUROUX (E.), "*Une méthode d'enquête anthropologique appliqué à l'Ouest malgache*", Paris, 152 p., December 2002.
13. FERREOL (G), *"Histoire de la pensée sociologique"*, published by Armand Colin, 1994
14. FOURNIER(J) et QUESTIAUX(N), " *Traité du social : situation, luttes, politiques, institutions* ", 4ème Edition, E. Dalloz 1984.p55-p645
15. FRAISSE (P) and PIAGET (J): *"Traité de psychologie expérimentale : Motivation, Émotion et Personnalité"*, Paris PUF, 1983.
16. GEORGE (P), *"Les pays tropicaux"*, PUF, Paris, 1968.
17. GEORGE (P), *"Questions de géographie de la population"*, Paris, PUF, 1959.
18. GEORGE(P), *"Sociologie et Géographie"*, Paris, PUF, 1968.
19. GUICHAOUA (A.) and GOUSSAULT (Y.), *"Sciences sociales et développement"*, Armand Colin, Paris.
20. GURVITCH (G.), *"Dialectique et sociologie"*, Flammarion, Paris, 1962.

21. GRANDIDIER (A.), "*Histoire physique, naturelle et politique*", Volume IV, Tome 4, Paris, Librairie Hachette, Société d'édition géographiques, maritimes et coloniales, 1928.

22. GRAWITZ(M), "*Méthode des sciences sociales*", Dalloz, 920 p, 1996.

23. ILUTCH (I.), "*Libérer l'avenir*", Paris seuil, 1971.

24. MANUEL de (Q.): "*Interactionnisme symbolique*", Rennes 2, PUF, 1994.

25. MARLAINE (C.) & Françoise (O): "*Sociologie de l'éducation*", Édition la découverte, Paris, 1995.

26. MARTINET(D), "*Culture prolétarienne*", Paris, Maspero, 1976.

27. MAUSS (M), "*Sociologie et Mythologie*", 4th Edition, in Bibliothèque de sociologie contemporaine, 1969.

28. PARSONS (T), "*La structure de l'action sociale*", Paris, Edition PUF, 1937.

29. PELLOUX (R), "*Le citoyen devant l'Etat*", Paris, Edition PUF, 1963.

30. POIRIER (J), "*Ethnologie et Développement*", in CRASOM, p. 207, 1972.

31. POULANTZAS (N.) "*Pouvoir politique et classes sociales*", Tomes II, Maspero, 1975.

32. QUICHAOUA (A) and GOUSSAUT (Y), "*Sciences sociales et développement*", Armand colin, Paris, 189 p.

33. QUIVY (R), VAN CAMPENHOUDT(L), "*Manuel de recherche en sciences sociales*" Dunod, Paris 2nd Edition, 287 p, 1995.

34. RANDRIANARISOA (P): "*L'enfant et son éducation dans la civilisation traditionnelle malgache*", Tome 1, SME, 1981.

35. RANDRIANARISON (J): "*Le Bœuf dans l'économie rurale à Madagascar*", Revue de géographie, volume 29; 81 p, Juillet-Décembre, 1976.

36. RASAMOELINA (H): "*Madagascar. Etat, Communautés villageoises et banditisme rurale*", Paris L'harmattan, 250p, 2007.

37. ROCHER (G): "*Introduction à la sociologie de la famille*" Paris 6, Points, Sciences Humaines, 1970.

38. ROUSSEAU (J.J), "*Du contrat social*" by BERTRAND De Jouvenel, Ed le livre de poche, Paris 1978,445 p

39. ROUVEYRAN (J.C), "*La sociologie des Agricultures de transition*", Maisonneuve et la Rose, Paris, 1972

40. SCWARTZENBERG (R.G), "*Sociologie politique*", Ed Paris, 1991,592 p

41. SEGALEN (M): "*Sociologie de la famille*", Paris, Armand Colin, 2004.

42. SILVANO (J) and CONTE (B): "*Le Changement social*", (Emile DURKHEIM), University of Lille, 2001.

43. TONNIES (F.): "*Communauté et Société*", Paris PUF, 1982.

44. TOURAINE (A.), *"Production de la société",* Edition le Seuil, Paris

45. WEBER (M.), *"Economie et société, tome 1: Les catégories de la sociologie",* Plon/Agora, Paris.

46. WOLF, *"Sociologie économique",* Edition Cujas, 1971.

SPECIAL WORKS

47. ALTHABE (G), "*Anthropologie politique d'une décolonisation"*, L'harmattan, 2000.

48. AMIN (S.), *"Le développement inégal",* Editions de Minuit, 1973.

49. AUJAC(H), *"Une hypothèse de travail : l'inflation ; Conséquence monétaire des comportements des groupes sociaux"* ; In Economie appliquée, t III, n°2, PUF Paris 1950.

50. BALANDIER (G.), *"L'Afrique ambigüe",* PUF, 1957.

51. BATTISTINI (R*), "Géomorphologie, Madagascar, Extrême-Sud", Toulouse,* Edition Cujas, Tome II, (1964), 638p.

52. BETTELHEIM (B)*, "Le cœur conscient",* Paris, Robert Laffont SA 1977 383p

53. BONTENS (H) and ROTILLON (G.), *"Economie de l'environnement",* La découverte p. 15, 1998

54. BURNEY (D.A), *"Theories and facts regarding Holocene environmental change before and after colonization. In Natural change and human impact in Madagascar",* S.M. Goodman & B.D. Patterson (eds.), Smithsonian Institution Press, Washington & London, Pp. 75-89, 1997.

55. CONDOMINAS (G), *"Fokonolona et collectivité rurale en Imerina"* Ed Berger Levrault, Paris 234 p, 1960.

56. CROIZIER (M) et FRIEDBERG (E), " *L'acteur et le système (les contraintes de l'action)* " édition du seuil, Paris France 478p, 1993

57. CROZIER (M), TROSA(S) (dir), *"la Décentralisation, Réforme de l'Etat",* Editions pouvoirs locaux, 1992

58. DECASTRO (J), *"Géographie de la faim",* Ed du Seuil, Paris, 1964.

59. DEMONQUE (M), EICHENDRERGER (J. Y), *"La participation",* France Empire, Paris, 228p. 1968

60. DEWEY (J.), "*Le Public et ses problèmes",* Paris, Publication de l'Université de Pau/Farrago, 2003.

61. FAUROUX (E.), "*Une étude pluridisciplinaire des sociétés pastorales de l'ensemble méridional de Madagascar"*. ORSTOM. Cah. Sci. Hum. 25 (4) 1989: 489-497, 1989

62. FORTIN (S.), *"La participation et le pouvoir"* in Vol. n1-2, Québec, Université de Laval, pp. 307-307, 1969.

63. FRANK (A. G.), *"Capitalism and underdevelopment in Latin America*",

Published by Maspero, 1968.

64. GENRO (T.) & UBIRATAN de (S.), *"Quand les habitants gèrent vraiment leur ville".* Edition Charles Léopold Mayer 35 rue St Sabin 75011 Paris

65. GOLTICH (J. P.), "*l'approche coopérative en milieu rural et urbain"*, 1992, p 67

66. GONTCHAROFF (G.), "*Dix territoires d'hier et d'aujourd'hui pour mieux comprendre le développement local"*, Adels, 2009

67. HUFTY (M.), RAZANAMANATSOA (A.) and CHOLLET (M.): *Green neo-colonialism in Madagascar?* Sabelli ed. 1995.

68. KAUFMANN (J.C.), TSIRAHAMBA (S.). *"Forests and Thorns: Conditions of Change Affecting Mahafale Pastoralists in Southwestern Madagascar".* Conservat. Soc, 4 :231-61, 2006

69. LEBIGRE (J.M.) and BELLERA (F.), *"La Transformation Récente Des Paysages Dans la Région d'Androka (pays Mahafale)"*. In: Milieux Et Sociétés Dans le Sud-Ouest De Madagascar. Collection Îles et Archipels, 23 (Ed. J.M. Lebigre), pp. 27-42. Centre de Recherche sur les Espaces Tropicaux (CRET), Bordeaux 3, 1997.

70. LECOMTE (J.B). "*L'aide par projet : Limites et Alternatives"*, Paris OECD, 147p, 1986.

71. MALDIDIER (C.), "*La décentralisation de la gestion des ressources renouvelables à Madagascar _ Les premiers enseignements sur les processus en cours et les méthodes d'intervention"*, 133p, March 2001.

72. MANZOORE (A), *"La participation communautaire, clef de voûte des soins de santé primaires",* in carnet de l'enfance, gouvernement et santé de peuples, vol 42, Autonome, 1978.

73. MARC (G), *"Le développement en quête d'acteurs"*, Paris, Centurion, 1984,163p

74. MARX (K.), *"Contribution à la critique de l'économie politique",* Editions Sociales, Paris, 1957.

75. MASLOW (A.), *"Devenir le meilleur de soi-même",* Nouveaux Horizons, Paris, 2011.

76. MEISTER (A.), "*La participation dans les associations",* Paris, Ed. Ouvrière; 276 pages, 1974

77. MONTAGNE (P.), RAZANAMAHARO (Z.), COOKE (A.): *"Le transfert de gestion à Madagascar dix ans d'efforts : TANTEZA"* / 207 pages, 2007.

78. PECQUEUR (B.), "*Le Développement local",* Syros, 2nd revised and expanded edition, 2000.

79. PAMARD (C.B.) et FAUROUX (E.), *"L'illusion participative- Exemples Ouest Malgaches"*, Autre part, n° 31, 2004/3.

80. PERROUX (F.), *"L'Economie du XXème siècle"*; Presse Universitaire de Grenoble, Grenoble p.383, 1991.

81. PRUDON(P), *"Médicaments pour l'an 2000"* Lausanne Ed, d'en bas, 216 P, 1983.

82. RABEARIMANANA(G) ; RAMAMONJISOA(J) et RAKOTO Ramiarantsoa (H) : " *Paysanneries malgaches dans la crise* ", Edition KARTHALA 1994. 379p

83. RAKOTO Ramiarantsoa (H) : " *Chair de la terre, œil de l'eau... Paysanneries et recompositions de campagne en Imerina (Madagascar)* " Editions de L'ORSTOM Paris 1995 ; 364 Pages

84. RAKOTO(I), RAMIANDRASOA (F), RANDRIAMBOAVONJY (R), " *Histoire des institutions* ", Musée d'art et d'archéologie, Antananarivo, 1995, 274p

85. RAMASINDRAIBE (P), *"Fokonolona Fototry ny Firenena"*, NIAG, Tana, 216 p, October 1973.

86. RANAIVO (C), *"Le Fokonolona"*, Madagascar cahier, Charles de Foucauld Paris 1950.

87. RANAIVOMANANA (L) & AMPILAHY (L): *"Contribution à l'étude d'impact environnemental à la gouvernance des aires marines protégées : de l'analyse de la problématique à la cohérence des outils"* ONE, 2010.

88. ROSSI (G.), *"L'ingérence écologique. Environnement et développement rural du nord au sud"*, Paris, 2000.

89. ROSTOW W.W, -The *Process of Economic* Growth. 1953.

90. SMITH (A.) quoted by MANDRARA (E. T.), 2005. *"Cours d'histoire de la pensé économique"*, Faculty of Law, Economics, Management and Sociology - University of Antananarivo.

91. VACHON (B.), *"Le développement local intégré : une approche humaniste, économique et écologique du développement des collectivités locales"* - Dinner conference, Carrefour de relance de l'économie et de l'emploi du centre de Québec et de Vanier Domaine Maizerets, Québec, 2001.

THESES AND DISSERTATIONS

92. FANNY (R.), *"Gestion des forêts sèches à Madagascar et au Niger. Vulnérabilité et fonctions des systèmes socio-écologiques pour comprendre les reformes forestières et leurs effets"*, PhD thesis, 2012

93. GUYOT (L.), *"Reconnaissance Hydrogéologique pour l'alimentation en eau d'une plaine littorale en milieu semi-aride : sud-ouest Madagascar"*, PhD thesis, 2002.

94. MUTTENZER (F.), *"Déforestation et droit coutumier à Madagascar : L'historicité d'une politique foncière"*, Doctoral thesis, 2006

95. PONGCHAI (D.), *"Interactions between cattle farming and reforestation in the upland socio-ecosystem of Nan province in northern Thailand: a companion modelling process to improve landscape management"*, PhD thesis, 2010.

96. RAZAKANDRENY (C.). *Cattle breeding in Ampanihy-Ouest*. DEA dissertation. University of Tuléar, 2005.

97. RAZAFIARIJAONA (J), *"Sols, problématiques foncières et développement rural à*

Madagascar : dimensions anthropoïde-juridiques des rapports fonciers-environnements", Enseignant Chercheur, Conference, FLSH and ESSA, University of Antananarivo, 6 p, 2000

98. VERIZA (R.F.), *"Stratégies pastorales et gestion des ressources naturelles aux environs du Parc National de Tsimanampesotse",* University of Toliara, 2005, 137p

99. RANDRIANARIVONY (D.E.), *"Pouvoir et symboles en pays Betsimisaraka : le Tangalamena",* Thesis for Doctorat nouveau régime in sociology. University of Antananarivo, 2014

MAGAZINES AND REPORTS:

100. WORLD BANK 2000-2001: *"Fighting Poverty".* World Development Report.

101. WORLD BANK: Centre for Information and Knowledge Management Africa Region: *"Indigenous Knowledge for Development"* 45 p, Nov. 1998.

102. Cahier de l'Institut d'Etude du développement, *"Le savoir et Le faire",* Geneva, le Cerf, 1979, 197 p.

103. COGESFOR, *"Projet Gestion durable des ressources naturelles pour la conservation de trois régions hot spot de la biodiversité à Madagascar",* FFEM Madagascar project presentation report, FFEM, 13 June 2008, 68p, 2008.

104. IMF (International Monetary Fund) *"Finance and Development: Helping the Poor",* quarterly publication, Dec. 2000.

105. GOEDEFROIT (S.), *"La part maudite des pêcheurs de crevette à Madagascar",* Etudes Rurales, 2002.

WEBOGRAPHY

106. Bertrand (Y.A.). *"Comprendre l'environnement et ses enjeux",* www.ifri.org/files/europe/compte rendu PAC.pdf, 2007. Accessed on 12 April 2015

107. KONE (M.), *"Approche socio anthropologique de la question foncière",* https://docs.google.com/viewer?a=v&q=cache:p2F112cR0TEJ:inst-ethnosociologie.net/ Approche.pdf+KONE+Mariatou,+2009,+Approche+socio+anthropologique+de+la+question+fonci%C3%A8re&hl=fr&gl=fr&pid=bl&srcid=ADGEESifwYcLNjgr6izw9OothlMzx9kB2IcwnkShSxdf92dkspeEdWKuwjO0IknBQUeIUm5mVL9cQ1h0VJ79lAijkPpyrx4u_REDp4X3M7LGDxQHumpYqjBoLP2ASwO04V5hBNqyCRc&&sig=AHIEtbQgAiVxx_kwaZdpWhdP4kvKYUsIRDA , updated 2009, accessed 4 February 2013.

108. NZE BEKALE (L.), "*Aspect de la démocratie participative en Afrique francophone. Esquisse de l'organisation juridique de la participation locale, au Benin, au Burkina Faso et au Mali".* Grenoble (France): Université Pierre Mendès France. Retrieved 22 July 2013 from http://www.med-eu.org/proceedings/MED2/.

TEXTS AND LAWS :

109. Constitution of the 4th Republic of Madagascar

110. Law no. 90-033 of 21 December 1990 on the Malagasy Environment Charter (*J.O. no. 2035 of 24.12.90, p. 2540*), amended by law no. 97-012 of 6 June 1997 (*J.O. of 09.06.97, p. 1171, Special Edition and no. 2584 of 12.07.99, p. 1479*).

111. Law no. 96-025 of 30 September 1996, known as the GELOSE law, relating to the local management of renewable natural resources, (*J.O.R.M., no. 2390 of 14 October 1996, p. 2385*).

112. Law no. 97-017 of 8 August 1997 revising forestry legislation, decree no. 2005-849 of 13 December 2005 (J.O.R.M. *no. 2449 of 25 August 1994, p. 1717*).

113. Law no. 2001-005 of 25 July 2001 on the management code for protected areas, promulgated on 05 February 2005.

114. Law no. 2005-019 of 17 October 2005 setting out the principles governing the status of land (*J.O.R.M. no. 3007 of 2 January 2006, pages 4 to 15*).

115. Law no. 2005-021 of 17 October 2005 on the Mining Code (J.O. no. 3015 of 20 February 2006, pages 1569 to 1597), amending law no. 99-022 of 19 August 1999 (*J.O.R.M. no. 2595 of 30 August 1999, pages 1978 et seq.*).

116. Law no. 2005-022 of 2 August 2005 amending certain provisions of law no. 2001-031 of 8 October 2002 establishing a special regime for major investments in the Malagasy mining sector (LGIM). (*J.O.R.M., n° 3015, 20/02/06, p. 1597*).

117. Law n° 2006-031, of 24 November 2006, establishing the legal regime for untitled private land ownership, (*J.O.R.M., n° 3089, of 26 February 2008, p-p. 1630 - 1631*)

118. Law no. 2008-013 of 23 July 2008 on the Public Domain (*Extract from J.O.R.M. no. 3217 of 20 October 2008*).

119. Law n° 2008-014, of 23 July 2008, on the Private Domain of the State, Decentralised Authorities and Legal Entities of Public Law, (*Extract from the J.O.R.M, n°3218, of 27 October 2008*).

120. Draft law n°028-2008, dated 29 October 2008, overhauling the Protected Areas Management Code.

121. Decree no. 98-610 of 13 August 1998, regulating the terms and conditions for the implementation of Relative Land Tenure Security, in application of law no. 90-012 of 6 June 1997 amending and supplementing law no. 90-033 of 21 October 1990 on the Environment Charter, (*J.O.R.M., no. 2035 of 24.12.90, p. 2540*).

122. Decree no. 99-954 of 15 December 1999, amended by Decree no. 2004-167 of 03 February 2004, on making investments compatible with the environment or MECIE (*published in Official Gazette no. 2648 of 10 July 2000 and no. 2904 of 24 May 2004*).

123. Decree no. 2000-027 of 13 January 2000, regulating local communities responsible for the local management of renewable natural resources.

124. Decree no. 2001-122 of 14 February 2001, setting out the conditions for implementing contract-based management of State forests.

125. Decree no. 2005-849 of 13 December 2005, revising the general conditions of application of law no. 97-017 of 08 August 1997 revising forestry legislation. (J.O. n° 3024 of 17/04/06, pages 2099 to 2106).

126. Decree no. 2007-1109 of 18 December 2007, implementing law no. 2006-031 of 24 November 2006, establishing the legal regime for untitled private land ownership.

127. Order n° 14-DIS/AMP/FKT, dated 09 November 2009, designating the Chiefs and Deputy Chiefs of *the* 16 rural Communes of the *Ampanihy* district.

128. Interministerial Order no. 52005/2010 of 20 December 2010, amending the Interministerial Mine-Forestry Order of 17 October 2008 placing the sites covered by the Order of 18 October 2006 under temporary global protection and lifting the suspension of the granting of mining and forestry permits for certain sites.

SUMMARY

The aim of this study is to assess the scope and limits of the Gélose law and the GCF decree using management transfers established on the outskirts of a protected area in the zone where drought prevails. In this context, the *Mahafale* plateau, home to the *Tsimanampesotse* National Park, was chosen to test three hypotheses on 25 management transfers: (i) the level of participation of the grassroots community depends on its level of education, and non-participants at the lower end of the scale do not want to get involved in matters they consider obscure and dangerous; or they are simply protecting their interests by exploiting renewable resources ; (ii) the stability and success of the management transfers studied are ensured by the habits and customs as well as the traditional social organisation of the *Mahafale* plateau; and (iii) the management transfers studied do not achieve the objectives of sustainable management because of the reduction in the access and use rights of the *Mahafale* communities as well as the geographical situation and climate change which weigh heavily on their land.

Through three successive stages, 8 months of discontinuous fieldwork yielded results that validated these hypotheses. Secondly, most of the forests on the *Mahafale* plateau are taboo or sacred forests managed by a spirit called *Tambahoaka* and the *ombiasy*, the rest are forests cleared for maize growing. These forests are currently being taken over by the local communities, and since the communities are becoming impoverished as a result of the drought and the non-application of the *dina*, the sustainability of the management transfers around the park is uncertain.

Keywords: *participation, conservation, landscape, Mahafale plateau, biodiversity, ecosystem, grassroots community, mahafalisation, Tsimanampesotse.*

Author's address: Lot VT85 HBJR Andohanimandroseza, tel. +261 33 12 318 77 / +261 34 41 174 1 6

E-mail: prambinizandry@yahoo.fr or prambinizandry@gmail.com

TABLE OF CONTENTS

MIX
Papier aus verantwortungsvollen Quellen
Paper from responsible sources
FSC® C105338

Printed by Books on Demand GmbH, Norderstedt / Germany